THE DIGITAL WORKFORCE

The 5-step methodology to smarter workforce management

JARROD MCGRATH

First published in March 2018 by Major Street Publishing Pty Ltd
Reprinted in October 2018 and May 2019
E: info@majorstreet.com.au
W: majorstreet.com.au
M: +61 421 707 983

PO Box 295, Wahroonga, NSW 2076
E: jm@smartwfm.com
W: smartwfm.com
M: +61 412 901 117

Ordering information

Quantity sales. Special discounts are available on quantity purchases by corporations, associations and others. We can also produce customised editions of this book branded with your own logo. For details, contact the author.

Individual sales. This book is available from online booksellers and can be ordered from www.smartwfm.com. It is also available as an ebook from your favourite ebookseller.

Orders for university textbook/course adoption use. For orders of this nature, please contact the author.

A catalogue record for this book is available from the National Library of Australia

ISBN: 978-0-6480875-6-4

Edited by Sally Asnicar, Full Proofreading Services
Internal design by Production Works
Cover design by Simone Geary
Printed in Australia by McPhersons Printing Group

10 9 8 7 6 5 4 3 2

PRAISE FOR JARROD MCGRATH

Scott Farquhar [Co-founder, Atlassian]

Jarrod McGrath and his organisation, Smart WFM are passionate members of Pledge 1% (pledge1percent.org) - a global movement that encourages and empowers companies of all sizes and stages to integrate giving back into their DNA. I applaud Jarrod for supporting the Cathy Freeman Foundation and the Pledge 1% movement with a portion of the proceeds from his new book *The Digital Workforce. The 5-step methodology to smarter workforce management.*

Jamie Howden [CEO, the Cathy Freeman Foundation]

'Thank you to Jarrod for his ongoing support of the Cathy Freeman Foundation and for embracing corporate philanthropy as a cornerstone of his business. Business leaders like Jarrod, who continually push the boundaries on giving, allow us to get on with the important task of bridging the education gap between Indigenous and non Indigenous Australians. I am particularly grateful to Jarrod, for committing a portion of all book sales proceeds to the Cathy Freeman Foundation and Pledge 1%.'

Georgegina Poulos [Global Director People, T2 Tea]

'Jarrod is extremely knowledgeable in all things workforce management, and yet he still craves the new and the unknown to continue to push the boundaries in this area.'

Rob Scott [Global Lead Strategy and Innovation, Presence of IT]

'Whenever I think about a WFM problem or opportunity, I can't help but think what Jarrod would advise. To me, his name is synonymous with leading WFM thinking and outcomes that maximise the value of people in the work environment.'

David Guazzarotto [Digital HR Strategist and CEO, Future Knowledge]

'The accelerated changes brought on by the Digital Age affect every aspect of work. Jarrod McGrath is at the forefront of leading this change, carefully crafting strategic insights with practical applications to help leaders steer their organization of the future.'

Cian McLoughlin [Bestselling author and CEO, Trinity Perspectives]

'The world of business is evolving at a rate of knots, as new technology, work practices and customer needs emerge. Jarrod's book provides a blueprint on how to navigate this brave new world, providing clear and actionable insights for business leaders to follow. Packed full of fascinating interviews, industry anecdotes and creative business strategies, this book is a must read for anyone who is serious about people management in their business.'

DEDICATION

To my brother-in-arms Josef Sikora. Josef has listened to my ranting and raving all these years since we were kids playing in the paddock; shooting the slug gun in the backyard; creating my first digital art business back in the late '90s; through to my latest ventures in workforce management (WFM) – and everything else in between. Thanks, brother.

To my wife Michelle for listening to me over the years and for your relentless support in making this book a reality. I love you. I'm sure you can't wait for me to write the next book ...

To my kids Molly, Tadhg and Billie Mae. And yes, I'll acknowledge our two rascal dogs Finn and Fudge who are sitting on the couch next to me as I write this.

Thanks to Mum, Dad and my sister Jacinta for always supporting me and giving me the confidence to do what I do.

Jason Averbook, I love the way you continually push to ensure people initiatives provide strategic value to business. I am delighted with your positioning of WFM and its importance to HR, payroll and ultimately whole-of-business strategy.

Huge thanks to the interviewees for the book, Georgegina Poulos, Aron Ain and Matthew Michalewicz. Your knowledge, experience and wisdom help make the Smart WFM 5-Step Methodology tangible.

To the teams, customers and suppliers I have worked with over the years. I appreciate your belief, support and trust.

To my publisher Lesley Williams and editor Sally Asnicar. I couldn't have got this book across the line without you.

Honourable mention to Scott Farquhar for getting me hooked up with Pledge 1% Foundation. And thanks to Cathy Freeman, Jamie Howden and Luke Toebelmann of the Cathy Freeman Foundation for bridging the education gap between Indigenous and non-Indigenous Australians. Being involved with both these foundations has been life-changing for me. Keep up the great work, and I encourage everyone to help others in any way you can.

A special dedication to those who are doing what they love because they want to provide a greater sense of purpose to those around them and to give back to the community.

Thanks to the libraries of Sydney for hosting me while I wrote this book. Having three young children (and their friends) at home meant that alternative writing locations were needed, else Dad would have been unproductive and grumpy.

I have a great love of music and when I was looking for inspiration I found myself constantly playing *Dream Your Life Away* by Vance Joy, *Melodrama* by Lorde and listening to Triple J. Thank you for the music!

Lastly, I was lucky to be introduced to T2's Singapore Breakfast Tea, and I drank many, many, many cups of this tea along my writing journey. Thank God for T2!

'Only 7 per cent of organisations manage their workforce effectively'[1]

I could not believe this figure when I read it. How could this measure be so low in 2017? We live in a time where information is plentiful and scrutiny is greater than ever before, yet we can't manage our workforce effectively?

I hope this book provides some practical insights and new, useful knowledge to help you manage your workforce effectively, resulting in happier, healthier employees and a brighter, more profitable future for your organisation.

1 Ventana Research, 'Workforce Management Value Index 2017, Vendor and Product Assessment', January, viewed 10 February 2017, <https://www.ventanaresearch.com/value_index/human_capital_management/workforce_management>.

CONTENTS

FOREWORD

In our ever-changing world, the ability to digitally manage the workforce of today is more *important* than ever. Not only is managing the workforce important, it now offers a critical competitive advantage to organisations worldwide.

The concept of capturing a worker's time and paying them accordingly is nothing new. We have been working too hard over the past century to accurately compensate individuals for the work they do. In a manufacturing economy this was tough enough and based on an individual showing up for work, documenting their attendance and processing their pay. The ability to do this has been a mainly manual effort and one that is prone to error as well as filled with manual heroics to simply ensure that the workforce gets paid. Well, the future of the workforce combined with the process of managing it has become much more complex, creating the need for our function to be more agile than ever. Welcome to the world of the 'digital workforce'.

In *The Digital Workforce*, Jarrod documents the journey we have been on for the past century from the second industrial revolution of the assembly line to where we are today in the fourth industrial revolution; a world filled with artificial intelligence and machine learning. During this time, our role has become more important as well as complex, shifting from simply capturing time and processing pay, to a world today that focuses on using data to forecast, predict and provide the business with tools never imagined until recently. This shift involves much more than technology; it truly touches the core of the business and creates a set of intelligence that will make or break the future of an organisation.

The Digital Workforce teaches us the important lessons on how true workforce management can solve problems. It provides an important lens through which all in HR and payroll and other lines of business can look to not just understand but mandate a new focus on the area of workforce management. Jarrod does a brilliant job of detailing the four key areas that workforce management impacts in a business by thoroughly documenting the *visibility, experience, flexibility* and *compliance* outcomes that true workforce management can provide. All organisations worldwide are constantly looking to show the return on investment (ROI) of their initiatives and, through Jarrod's work on documenting the problems that workforce management can solve, we now have the toolkit we need to show not only ROI but true VOI, (value of our investment) from our advanced and enhanced practices.

We all have learned in this industry that proving and showing value is one thing but taking action is another. *The Digital Workforce* not only helps us with value, it details the step-by-step processes required to achieve ultimate success. In order to achieve success, this book walks us through a five-step process that discusses *alignment* to *preparation*; from *implementation* to *tracking our results*; and finally it teaches us how to not only 'go live' but truly *measure* the impact of our efforts. This five-step process has become and will continue to be the bible on how to lead any workforce management initiative. In each step of this SMART process, *The Digital Workforce* gives us examples, learnings, actions and takeaways that are 'musts' in our workforce management initiatives. The other invaluable way Jarrod documents this process is through the examples he provides, ranging from industry to type of workforce being managed to the different technological solutions to achieve results. Again, proving value is one thing, but providing the proof is another and Jarrod has done this with great success.

The successful deployment of workforce management as a practice requires a careful blend of people, process and technology. *The Digital Workforce* does a comprehensive job of combining these concepts and sharing examples and requirements of the people needed for success, the processes that need to be digitised and 'reimagined' and finally he goes into depth discussing the technologies available in the market today and what impact the world of artificial intelligence and robotics will have on our industry. Never in our history have we had a documented source like *The Digital Workforce* that can truly help us move from concept to context to proof of success.

The Digital Workforce is a must-read for anyone looking to capture the value of their workforce from a business impact point of view, as well as for all of us looking to make our industry one that shifts from transactional to strategic. As someone who has spent their career in the HR and workforce technology industry, working with leaders around the world creating strategies and actions in alignment with their visions, I truly appreciate how hard this work is and the effort that goes into it. The fact that Jarrod has documented these efforts in a manner that is easy to comprehend gives us a tool as an industry that will create value and educate us all for decades to come.

Congratulations Jarrod on an amazing piece of work. I know readers will enjoy it as much as I have and will for years to come.

Jason Averbook
CEO, Leapgen

Jason Averbook is a leading analyst, thought leader and consultant in the area of human resources, the future of work and the impact technology has on that future. He is the Co-Founder and CEO of Leapgen, a global consultancy helping organisations shape their future workplace by adopting forward-looking workforce practices and fast-innovating technologies personalised for their business. Jason has more than 20 years' experience in the HR and technology industries and has collaborated with industry-leading companies in transforming their HR organisations into strategic partners.

ABOUT THE AUTHOR

Jarrod McGrath grew up in the small country town of Bathurst in New South Wales, Australia. Bathurst is famous for three things: farming, motor car racing and education. As a young bloke riding motorbikes around on farms, it became obvious to Jarrod that agriculture was not his calling. He fell in love with cars, especially fast ones, which is still a passion of his - but it's not the topic of this book!

Jarrod's father and other relatives and friends are educators, school teachers and volunteers. This taught him at an early age that education is a fundamental foundation stone on which you can build a lifetime of learning. It's also how you learn to challenge and shape what the future holds. Jarrod loves to learn and to challenge - and hopefully shape - what the future may bring.

Jarrod studied the theory of workforce management (WFM) and artificial intelligence (AI) at university. WFM is founded on an area of mathematics known as Operations Research, on which Jarrod focused in his degree. From the outset, Jarrod was intrigued to understand how a business could run more effectively and deliver greater value by optimal use of its assets. When he started working with WFM software in the early 2000s, this idea became tangible when he discovered that Operations Research could be applied to optimising employee productivity and increasing value.

Jarrod has dedicated a substantial part of his career to building the global profile of WFM, and over the years he has had significant hands-on WFM delivery experience from the perspective of the client, service-provider and product-supplier. He worked with enterprise resource planning (ERP) software in the late 1990s and had strong exposure to large enterprise implementations working for a major systems integrator and its clients. In more recent years, he founded and established a global WFM practice for a world-leading human capital management (HCM) player. This gave him more hands-on exposure to global WFM implementations across human resources (HR) and payroll.

In short, Jarrod has been responsible for, and successfully delivered, literally hundreds of WFM programs over the years to clients across a multitude of industries including retail, health, construction, manufacturing, hospitality, service and government. His expertise is in both local and international sales and delivery for large and small-to-medium-sized clients, solving an amazing number of interesting and valuable workforce-related business problems.

In his current role as the founder and CEO of Smart WFM (smartwfm.com), a boutique consultancy focused on empowering the workforce now and into the future, he focuses on what clients need to stay relevant in a time of rapid digital advancement.

In short, Jarrod McGrath is a visionary WFM leader who can articulate the strategic value of WFM within a business, from the senior management level through to the operational coal face. Hopefully, you find his book informative and useful on your own WFM journey. This industry is so fast-moving readers should regard this book as the first edition as Jarrod will soon need to update it!

Jarrod McGrath is a sought-after public speaker. For further information and to find out about Jarrod's availability, please contact: **speaking@smartwfm.com**

PREFACE

Workforce management (WFM) is about managing people-processes more efficiently, improving productivity, providing financial acuity, purpose and inspiration, building brand loyalty and enabling a digital workforce using all the technologies that we are spoilt with in the modern working environment.

The history of WFM dates back to the 1890s, and its evolution has been closely aligned with revolution - industrial revolution, that is. Few people realise there have actually been four industrial revolutions - the latest of which we are in right now.

Recently I read *The Fourth Industrial Revolution* by Klaus Schwab, Founder and Executive Chairman of the World Economic Forum. I was fascinated by his insights into the exponential growth of technology and the confluence of technologies such as artificial intelligence (AI), robotics, the internet of things (IoT), 3D printing, etc. The influence of technology is spread across nations, governments, economies, businesses and communities. The speed at which these technologies are impacting people and organisations is greater than ever before. Organisations and their workforces are being reshaped, realigned and re-tooled, as well as being challenged to ensure we keep a human-centred approach rather than dehumanising work processes.

There are strong correlations between the history of WFM and the second, third and fourth industrial revolutions, which I'll explain further in the first chapter. Briefly, the parallels are around the significance of the developments and our ability to adopt, adapt to and embrace them in a thoroughly ubiquitous way. Each of the industrial revolutions resulted in modernisation and transformation: the invention of steam engines; the construction of railways; the advent of electricity; the introduction of the assembly line; the development of the microprocessor; and machine learning, to name a few.

Over time, therefore, WFM has evolved from being a system of record, to a system of productivity due to these industrial advances.

Organisations today are fortunate in that they no longer have to accept things as being just 'the way they are'. Digital technologies - mobile devices, social media, job boards, knowledge portals, robotics, 3D printing all wrapped with smart computing and AI - allow us to redefine the way we work and network with anyone, anywhere, in real time. We can learn online and think for ourselves far more easily than ever before allowing us to challenge the norm and reset the status quo. Most importantly, digital technology gives organisations the opportunity to bring ideas together; senior management can drive the business across all areas, right down to the operational coal face. If we embrace digital technology, we can develop a positive, progressive digital workforce for the organisations of today and into the future.

Digital technologies are integral to business and indispensable in running a business effectively and competitively. Everything we do now has a digital flavour, the way we work has a digital flavour and the workforce transacts in digital. Digital is ubiquitous and enables 'the digital workforce' - hence the title of this book.

I was inspired to write this book to achieve three goals:

1. To teach organisations what WFM is by examining the various areas of WFM through a business-focused lens.
2. To make organisations aware of key, predictable WFM-related issues that arise.
3. To present my proven **5-Step Smart WFM Methodology** that can be used as a template to overcome WFM-related issues and deliver long-term, effective workforce value.

Very little has been written about what WFM is and, more importantly, how businesses can leverage its value. This book does both, but its main purpose is to teach you how WFM can benefit your business now and in the long term.

About this book

The Digital Workforce is a how-to book for senior managers, project sponsors and project teams tasked with transforming their workforce in a disruptive, digital era. There are plenty of anecdotes, real-life experiences and examples to help explain the concepts that I introduce.

I present the information in two parts.

Part I – What is workforce management?

In the first part, I answer three key questions:

1. What is WFM?
2. What does WFM mean for your people and your customers?
3. How does digital and AI impact your workforce?

To help put the answers to these key questions in context, I include three fascinating interviews with industry leaders who offer three different perspectives, that of:

→ **The vendor:** Aron Ain, Global CEO of Kronos Incorporated, provides his perspective on the history of WFM and where it's heading.

→ **The customer:** Georgegina (Georgi) Poulos, Global Director People, T2 Tea, speaks about WFM and its benefits to her business. Georgi explains the strategic importance of WFM, how digital has accelerated time-to-value and the importance of storytelling to drive change initiatives related to the adoption of WFM.

→ **The futurist:** The final interview is with world-leading AI expert Matthew Michalewicz who speaks about the rise of AI technology and its likely impact on people and workforces.

Part I concludes with some insights into where I see WFM heading in the next few years.

Part II – The 5-step methodology to smarter workforce management

In Part II, I present the Smart WFM 5-Step Methodology and how it can be used to get WFM right in your business.

Having overseen many WFM implementation programs over the years, I have learned that the process of implementation is predictable. So why do so few WFM programs meet their business objectives? I discovered that various business areas, particularly HR and payroll, have differing views on how to achieve value from the workforce; and there is often a lack of pertinent, cohesive information to make an informed business decision and tie this back to the organisation's objectives. You only get the full picture when you speak to everyone who can influence or contribute to a business outcome: leaders from HR, finance, IT, operations and payroll.

I have also observed that there are many vendor-led implementation methodologies that concentrate on successful implementation of the technology layer, but they don't focus on the changes required to adopt this successfully into their client's business.

I have developed the Smart WFM 5-Step Methodology from my experience of numerous business implementation programs and wrapped these experiences into a methodology that looks at success through your eyes, as well as the eyes of the business.

Here's what makes the Smart WFM 5-Step Methodology different from other methodologies on the market. The Smart WFM's 5-step methodology:

- → is relevant to WFM *today*, and it includes the impacts of digital and where the workforce of the future is moving.
- → can be used to support *any* phase of the WFM investment journey, from the beginning – alignment – to the final step – measurement – and everything in between.
- → is designed to be used in conjunction with vendor-led implementation methodologies and industry-recognised project management methodologies such as PMBOK and PRINCE2.[1]

- → is based on real-life experiences that represent a 360-degree view of the WFM market.

I encourage you to read *The Digital Workforce* from cover to cover, to learn more about WFM and how to leverage its value in your business. Alternatively, if you are looking for specifics, you can read relevant sections and take out nuggets of information to help with your particular area of interest.

You will find some topics are covered more than once in this book, to reinforce and emphasise their importance, and to cater for readers who may 'dip in and dip out'.

I've enjoyed the opportunities WFM has provided to me both professionally and personally over the years, and I'm delighted and privileged to have been able to take the time out to write this book for you. I hope you find *The Digital Workforce* enjoyable and useful to drive WFM value into your business. Furthermore, I hope this book helps to broaden your horizons and make people-focused activities both within the workplace and the community more productive and enjoyable, improving the experience for all concerned.

Challenge yourself and don't accept the norm because it's 'the way it is'. Be curious, think digital and think to the future of work.

All the best

Jarrod

1 Perhaps the two most common methodologies to deliver IT projects.

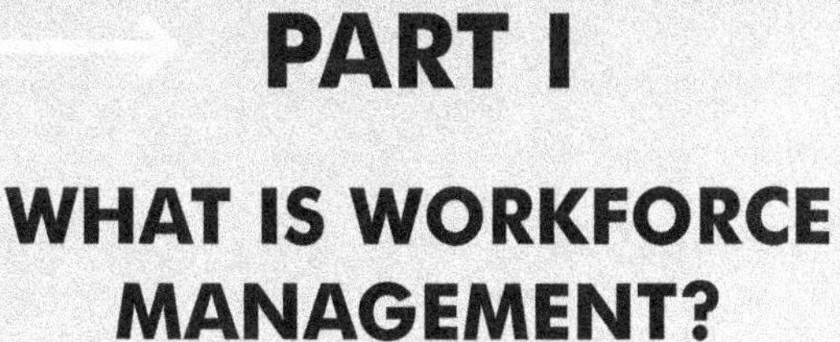

PART I

WHAT IS WORKFORCE MANAGEMENT?

'Clients do not come first. Employees come first.
If you take care of your employees,
they will take care of the client.'

Richard Branson

Chapter 1

THE EVOLUTION OF WORKFORCE MANAGEMENT

I was inspired to write this book because I realised 'workforce management' is a term many people have heard about but few can give a clear definition of; moreover, most do not fully understand the benefits workforce management (WFM) can bring.

When I first started working in WFM, I would have defined it as 'right person; right place; right time', and at the time this definition would have been relatively accurate. In the early 2000s, I would have extended the definition to 'right person; right place; right time; right skills'. Today, the areas that WFM affects and its benefits go far beyond this.

I recently facilitated a workshop at a prominent conference for HR and payroll professionals, where I posed the question 'What is workforce management?' The responses I received were many and varied:

- award interpretation
- payroll and costing
- work order management
- mobility
- Employee Self Service (ESS) and Manager Self Service (MSS)
- planning
- performance management.

This list of suggestions kept growing, but what stood out for me was that all the answers were different to mine. I also noticed that the definitions given were highly functional and transactional in nature – and lacked a common understanding of what WFM is.

So, here's my one-line, nutshell definition of WFM, and the following chapters will give examples that support this definition:

> *'WFM maximises people value, productivity and experience.'*

WFM is born of industrial revolution

To really understand what WFM is, we need to go back to the late 19th century and look at its evolution. Below is a brief chronicle of how WFM developed in line with three waves of industrial revolution over the past 100+ years; how its ability to problem-solve impacted various areas of organisations; and the spread of WFM from its relevance to a single industry, to benefiting all industries and professions.

1890s–1960s. The Bundy Clock

On 26 May 1891, Willard LeGrand Bundy patented the Bundy Clock.[1] Known as the 'Workman's Time Recorder', the invention of the Bundy Clock coincided with the second industrial revolution, a period of rapid industrialisation from the late 1800s to the early 20th century. At a time when most labour was employed in mass production and on assembly lines, the clock was used to record workers' start and stop times. The Bundy Clock thus increased the accuracy of hours worked and was, in effect, an early form of WFM.

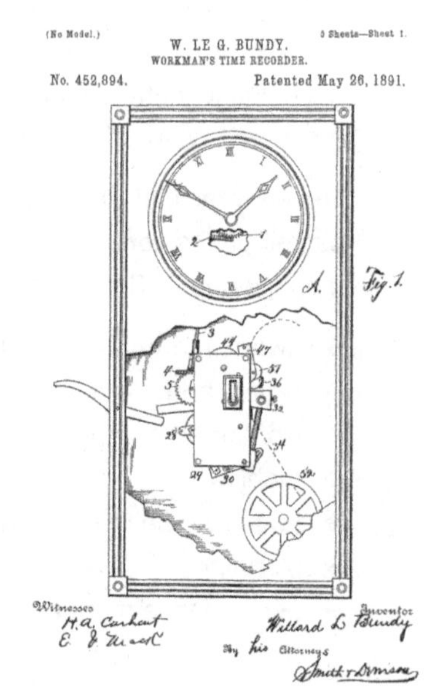

Over time, these records were fed to payroll to increase the accuracy of workers' pay. So from the 1890s to 1960s, payroll and staff were the primary areas affected by the invention of the Bundy Clock, and manufacturing was the main industry to benefit.

1970s–2000s

The last three decades of the 20th century and the early 21st century brought the third industrial revolution, which saw the invention of an 'online time clock' to coincide with the development of the microprocessor and the internet.

The first microprocessor-based time clock was developed in the late 1970s. This device enabled the digital automation of time capture. Functions of these devices gradually expanded to include self-service 'kiosk' functions, such as checking of leave balance, applying for leave, applying for a call-back allowance.

Due to the online nature of these solutions, managers were able to see in real time who was at work; what shifts their team members were rostered to; whether their team members had the right skills; the cost of each person on any given shift; how much overtime had been incurred; and over-coverage and under-coverage of shifts. In some industries, such as health, retail, hospitality and contact centres, where staffing was impacted by demand, managers could see the impacts of expected demand to forecast required staffing levels.

In this period of WFM evolution, some organisations elected to educate their supervisors to step up and become leaders and provided them with the required training to develop their leadership skills. This delivered the tools to have a complete view of their workforce, including skills and skill gaps. These leaders could then work proactively with their team members and HR to focus on value-add areas such as staff development.

From the 1970s to 2000s, the impacts of WFM had expanded from payroll and workers, to include supervisors, information technology (IT), finance and HR. In addition, industries now affected by WFM had expanded to include those where WFM could add value, such as health, retail, hospitality and contact centres.

2010s forward

The rapid development of digital technology, including mobility, tablets and wearables, and the development of artificial intelligence (AI), IoT, omni channel and machine-learning algorithms, coincided with our current period of advancement: the fourth industrial revolution.

Technology suppliers are developing technology at such a rapid rate that it is difficult for organisations, policymakers and governments – let alone employees and customers – to keep up.

Likewise, WFM has evolved to provide intelligent solutions in real time, such as being able to analyse staff attendance data to predict which days an employee is likely to be sick. WFM technology is able to trigger performance management events based on the number of employees who are late to work (late-ins as we call them). Wearables provide hardware that can automatically register when a worker has arrived at work and when they have left work for the day. When a rostered worker is sick, WFM technology will automatically find a suitable replacement worker for their shift without any human intervention.

Since the 2010s, the impact of WFM has expanded to include operations management, senior management and workforce robots. 'Workforce robot' is a term to define a robot that replaces a human role, but that role would still need to be filled by a human if the robot was not there.

In summary, WFM now covers all industries, with digital wrapped around it to drive greater value.

So, what does this mean for you and your organisation?

Your organisation now has a much greater level of workforce information available, enabling it to make the workforce more productive and improve its experiences, giving you a *complete* picture of your people.

The fourth industrial revolution and WFM

Why do I draw a correlation between industrial revolution and the evolution of WFM? Throughout history, each industrial revolutions has brought momentous change, and advances in WFM have been just as significant. As noted by Klaus Schwab,[2] two challenges face the fourth industrial revolution, which might also limit the potential of WFM:

1. Changes in leadership might be necessary to facilitate the large-scale changes across economic, social and political systems.
2. Lack of narrative outlining the opportunities and challenges: if we are to empower a diverse set of individuals and communities we need this narrative.

In my view, WFM is no different. If business does not adapt to workforce change, and in a way that people can relate to and understand, the benefits of WFM will not be realised. This is clearly demonstrated in the recent Ventana[3] survey, which found that only 7 per cent of organisations manage their workforce effectively.

A summary of the evolution if WFM, its impacts on people and industries, and its correlation to the second, third and fourth industrial revolutions is depicted in Figure 1.1 below.

Figure 1.1: The evolution of WFM

Start/stop accuracy, payroll	**Scheduling, automated pay real time, kiosk**	**Forecasting, prediction flexibility, AI**
Impact Payroll, staff	**Impact** Payroll, staff + Supervisors, IT, finance, HR	**Impact** Payroll, staff, supervisors, IT, finance, HR + Operations management, senior management, workforce robots
1890s – 1960s	**1970s – 2000s**	**2010s +**
Second industrial revolution Mass production, electricity and assembly line	**Third industrial revolution** Semiconductors, mainframe and personal computing, internet	**Fourth industrial revolution** Ubiquitous computing mobility, AI, machine learning

Interview: The Vendor – Aron Ain, CEO Kronos Incorporated

In the preface, I mentioned I will be examining three key areas of WFM through a business lens. We've looked at the first key area; now we'll look at WFM from a vendor perspective. It's my pleasure to include the following interview with Aron Ain, CEO of Kronos Incorporated. Kronos was one of the early suppliers of WFM technology and has been a pioneer in this space for the last 40 years. Here Aron speaks about how WFM has evolved and where it's heading.

About Kronos

Kronos is a privately held company founded in 1977. Headquartered in Lowell, Massachusetts, Kronos employs more than 5,000 people worldwide.

Kronos is a leading provider of workforce management and human capital management cloud solutions. Kronos's industry-centric workforce applications are purpose-built for commercial businesses, healthcare providers, educational institutions and government agencies of all sizes.

About Aron Ain – CEO, Kronos Incorporated

Aron Ain is Chief Executive Officer of Kronos and a member of the board of directors. Since he joined Kronos in 1979, Ain has played a role in nearly every functional department at the company, helping to build what is today a $1.3 billion global software enterprise.

Jarrod: Thanks, Aron, for taking the time to speak with me and share some of your knowledge. First, can you tell me a little bit about your career and how you got into the world of workforce management?

Aron: Sure. I've been in the world of workforce management a long time. I started in it before it was called workforce management. My career started with joining a small start-up company called Kronos, which was in the business of taking that everyday business practice – the time clock – which hadn't changed since the 1950s and automating it. That's how

I got into it. Whoever imagined it would grow to be what it is? Or the fact it would be called workforce management – but it is. In fact, I was involved with creating the term 'workforce management' back in 1982. I can show you the actual communication booklet we created that introduced the idea of workforce management to the world.

We decided to coin the phrase workforce management because when people thought about this technology-centric solution that was trying to automate this electro-mechanical world of mechanical time clocks, they immediately oriented themselves to think about it just like what a time clock did, as opposed to all the dynamics of what a workforce management system could do. Our attempt to change the way people were thinking about the opportunity was by starting a discipline called workforce management.

Jarrod: You've seen an amazing evolution of workforce management since that time. Can you describe some of the key milestones you've seen along the way, particularly since you joined Kronos?

Aron: I joined in 1979. In '82 we introduced the idea of workforce management to the world. As I mentioned, the mechanical time clock had always been the front-end of the payroll processing function. While payroll had been automated, if you will, gross-to-net calculations and automating the process of printing pay cheques were at the front-end of that process and were still done manually with organisations using time clocks or timesheets. What we did was we created a device that looked like a mechanical time clock, but it wasn't a mechanical time clock. It had a microprocessor in it. Instead of just telling you when people came and went, the microprocessor gave us the ability to program work rules in the device, so it would add up the punches. It would add up the registrations and apply the work rules, the awards, the collective bargaining agreements, according to how individual companies did that.

Now, that was exciting, but the problem was in US terms, those devices were expensive. They cost about $5,000 a device. We had a $5,000 solution to a $2,000 problem.

One of many turning points was in 1981 when IBM released the PC. That's, in fact, when it was announced. We were able to take all the intelligence of this expensive device that had a microprocessor in it, and, if you will, dumb it down to just become a data collection device. Then we put all the intelligence in this micro-computer, this personal computer. That was a major turning point. Then, that opened up the possibilities to really expand all components of the software that was in this system, so that we could then move past just collecting when people came and went to pay them accurately, but also do all the other things that now are key and integral parts of workforce management. So, not only keeping track of when people come and go, but keeping track of what they do while they're there, scheduling when they should be there, etc. This was an everyday business practice that managers were doing manually, and it's all related to workforce management.

Then, the world moved from PC to local area networks. It moved from local area networks to wide area networks. It moved from there to client servers. It moved from there to the Web. It moved from the Web in its early forms to the Web of today, and now it's moved to mobile, and social and artificial intelligence - all still trying to attack the same problem of how to effectively manage a workforce.

Jarrod: If you look back to when you were talking about time collection way back in the '70s, early '80s, I think as WFM has expanded, it has certainly cut across more people in an organisation now that it is across mobile, social and AI. How do you think that impact on those different departments and different functions has played out?

Aron: Well, what used to happen was it was the function of the payroll department to keep track of when people came and went. As workforce management expanded, it became much more a function of the front-line managers who were responsible for making sure the right person was in the right place at the right time, making sure the employees were engaged and focused in all those dimensions. It expanded from a payroll-centric function to making sure people worked accurately to a much more operational focus.

From my point of view, I'm always surprised, to some extent, when workforce management is included in the greater HCM [human capital management], HR stack. Other parts of HCM, payroll and benefits administration and employee self-service, and recruiting and compensation and performance and onboarding, they're more back office functions to some extent. They're run by centralised groups;

where what workforce management is, it is an operational function. It is right down at the centre of where activity and services are being delivered and products are being built, etc. To some extent, it's a sibling to HCM, but it has been grouped in broader HCM.

Jarrod: Yes, it absolutely has. In terms of the customers, when I embark on a workforce management program, what do you think - and I ask this question in the context of today - what are they really looking for?

Aron: They're looking for something on one dimension that can save costs and make sure they're in compliance and help drive productivity and those types of things, but I think, more importantly, as each day goes by, there's a deeper focus on driving business outcomes, delivering better service, driving employee engagement. Those are things that people never thought about before.

For example, if you're a retailer, and you can make sure you have the right person with the right skills in the right place at the right time, you will introduce a better customer experience; they will buy more from you, and it will impact your business outcomes - and there are examples of that.

If you're a hospital and you deliver better patient care by, again, in this example, having the right person in the right place with the right skill at the right time, then you will enhance your relationship with your customer, in this case, a patient, but still a client who has choices of where they go. People could choose to go to different medical centres to get their healthcare provided.

If you're a manufacturer who's building a product and you can do it more effectively, more efficiently, and drive higher quality, your customer will benefit from that and you will drive [better] business outcomes.

I think it's moved past this whole idea of cost and compliance to really focus now on these business outcomes around driving more revenue, delivering better service, higher customer engagement. Think about an employee who tells you what their optimal time to work is, and you can help them make sure they get there.

Jarrod: Are you seeing more businesses start to measure those outcomes as well as business outcomes?

Aron: I would say beyond *more* doing it, they're *all* doing it today. It's rare today, when we have engagements, that business outcomes aren't a key deciding factor in terms of what type of workforce management system [the customer is looking for]. It's become a key component of it. Yes, it's already crossed the chasm. It's there.

Jarrod: What is your advice to a customer who is embarking on a workforce management initiative?

Aron: Well, I think you need to start with understanding that there's a community of users who can benefit within your own organisation from a modern workforce management solution. Look beyond just the obvious benefits to those areas where business outcomes can really be impacted by having a solution that will meet your needs, broadly defined. I think the days of just buying something that will automate the calculations or the awards' interpretation are over and organisations are looking for more. Now, what that means is organisations are embedding analytics into their tools and they're finding ways to use artificial intelligence and machine learning to drive automation in things that people never imagined was possible, but it's possible today.

For example, you no longer have to require every manager within the organisation to approve time off by touching each request. You can have the system do that for you according to your own rules, your own guidelines. You need to make sure that you understand that the employees of today and the managers of today expect access to information; they expect to be able to use devices and to interact, to communicate with the company about their own particular needs. [You need to understand] this whole gig economy where everyone views themselves as being their own boss, even when they work at a place ... To the extent that you understand this, you can provide solutions that broadly meet this new definition of work and you will be better equipped to recruit great people, retain great people, deliver great products, drive business outcomes. You have to look at it in the broadest sense today.

Jarrod: Moving away from the customer to the vendor landscape; over recent years, the vendor landscape has become increasingly crowded. How do you see the landscape evolving and what do you see are the implications from a customer perspective around that?

Aron: Well, I think an easy answer is to go to the HCM providers and think that they can provide a full suite of solutions around workforce management. But, I really think a different skillset is required. It's a different world. It's why organisations sit independently as workforce management organisations, and quite frankly, in some senses workforce management organisations are as large as the full suite of the global HCM organisations in terms of what their revenues and profitability are. I think that the future, on one dimension, competitively is going to stay crowded, but I think there are only going to be a few choices of who can really meet the needs of global companies that want a consistent system that works around the world. In the world of workforce management, within countries, rules are different; across borders rules are different; across regions they're different - and it's very difficult to do this.

I also think that companies have learned that doing things where everything that you buy needs to be customised, then you're stuck on an island, is a 1980s' view of how to buy a system. I think today, the providers who can offer a way for customers and users to be able to grow with the system as technology changes and the world changes is certainly important.

Obviously, it goes without saying that if you're not providing a cloud-based system that can offer all the advantages that a cloud-based system can offer, you're not offering your customers a complete modern way to help manage and direct the system. The vendor needs to accept much of the responsibility to manage the application that's being run for any particular customer.

Then, vendors will have to be really well versed in the whole digital world - mobile and tablets, the whole world of artificial intelligence and machine learning - going forward. They will have to know how to live in this world and offer solutions that will save steps. I'm not sure everyone's going to keep up with that as well, but I still think there'll be lots of opportunities to buy from different people. You have to decide what your objective is.

Jarrod: I think a lot of the existing criteria for buying have been around for as long as you and I have. What else do you think needs to be taken into consideration in a digital world?

Aron: I think the core components are there, but I really think they've expanded, Jarrod. I don't think people were talking about using a workforce management solution to increase revenue. I don't think they were thinking about a workforce management solution to drive customer satisfaction and employee engagement. The workforce management solution was always there to make sure you're in compliance, to save costs, to improve productivity, but these new dimensions are the ones where the real benefits are, where the real rubber meets the road, where it's going to drive the outcomes that organisations are looking for.

Jarrod: And that's where it really cuts across so many of those different areas of the business – way outside of just payroll and operations. It now drops into marketing, it drops into sales, it drops into HR. It drops into so many other areas of the business.

Aron: That's the point, you know. We have customers who tell us that because we put the right person in the right place at the right time with the right skill – a retailer, for example – that the experience their customers tell them about when they walk into the store has changed dramatically, that the sales have gone up. Could you imagine? Who would have ever thought buying a workforce management solution would make it so your sales would increase?

Jarrod: A wonderful opportunity for those customers to go down a WFM path.

Aron: Right.

Jarrod: Two questions just to close the interview, Aron. There's a great social consideration now within organisations. How are you seeing those social considerations play out, and how do you see them impacting the future of work?

Aron: They're integral. In this gig economy employees have choices today, and they're very smart in terms of finding if they're unhappy in their job. It's not like the old days where they would say, 'Where could I go instead?' They can just go online now and find lots of opportunities where they can go instead. Social is at the centre of how people expect to be communicated with, how they expect to communicate back to us, and so it's going to become the centre of activity in terms of how they're interacting with a workforce management system. Today, we develop everything on the mobile device first, and it's all responsive. There's no such thing as a separate mobile strategy. Today, your strategy *is* mobile. It has to be there. It's just not these mobile devices, although they're a big part of how people interact on the social way.

I was at a presentation recently, where I discovered that on average, people who are 18 to 30 years old – and they just used this demographic as an example – spend two hours on average every day looking at their mobile device. Two hours. That's two hours they used to spend doing something else. You want to know what they're looking at? They're texting. They're on Facebook. They're watching videos. They're doing all these things that social is a big part of. It's not the only thing they're doing, but it's a major part of it. That's their centre of engagement. You know, when my phone rings at home, I don't even get up from the couch anymore to answer it because anyone who really wants to get a hold of me, will call me on my cell phone or text me. I know it's not somebody I want to talk to, most likely, when that happens. [Social] has become an integral part of everything we do, everything we do. We have to accept it and grasp it and embrace it and make it be our friend, which we're doing.

Jarrod: On the other side of social as well, from an organisational perspective, from a 'giving' social conscience perspective in terms of communities and organisations having that greater sense of purpose for their employees, what are you seeing in that respect at the moment, Aron?

Aron: Well, I'm not sure that connects directly to workforce management per se, but certainly, I can speak from my experiences with our company and my children. They want to work for a place that makes a difference, not just inside the company walls, but also in the broader community. Their own expectations are they want to make a difference in their communities and the world. Social tools make that much easier to do. It's impacting philanthropy, it's impacting volunteerism, it's impacting how people spend their time.

Jarrod: It is a little bit away from workforce management, but it's very much associated with the workforce. What I'm seeing is that employees and staff are actually wanting an organisation to have something that's driving that greater good, that greater sense of purpose. I'm seeing that it's also making those employees more loyal and responsive to the organisations that are taking that point of view.

Aron: Couldn't agree more. It's why at our company, we have a GiveInspired program, which supports that. And it's not just that; it's beyond just giving. It's also volunteering and encouraging that to happen. I know it makes us an attractive employer. Look, people join companies because of the company. They also decide to stay at the company because of the experience they have while they're there. You have to work in both dimensions around your brand.

Jarrod: Absolutely. The final question is a broad one. What is your greatest hope and your greatest fear in relation to workforce management?

Aron: Greatest hope? I can't think of a greatest hope from that point of view. I don't really think about it in that way. Obviously, being on the vendor's side, I always worry that we're going to miss some major shift in technology. We're going to miss some significant change that's taking place. We're just always focused on making sure that we stay aligned in these areas, around staying up on technology. We talked about some of it. Mobile's really big today; we need to stay with mobile. Artificial intelligence is really fully embedded in the world we live in today; we need to do that. We need to be responsive and help companies meet their particular needs and use that technology and those tools to make that possible.

Jarrod: That covers it from my perspective. Is there anything you'd like to add before we conclude?

Aron: No, I think you really sent a list of comprehensive questions and focus, so I think I'm good. Sound like it's going to be a great book and I can't wait to read it.

Jarrod: Thank you, Aron.

* * * * * *

Top take outs

- The term 'workforce management' was coined by Aron Ain in the early 1980s to expand people's thinking past recording the time employees started and finished their working day.
- The evolution of WFM has resulted in many more benefits being available to organisations today than in the past. We have moved from systems of record to systems of productivity.
- It is important to understand the evolution of WFM is tied to the second, third and fourth industrial revolutions, and that often significant change is required before companies can adopt its benefits.
- WFM cuts across most major functions in your organisation.
- My one-line nutshell definition of WFM is: maximising people value, productivity and experience.

Where to next?

I hope you enjoyed this informative background to WFM and its evolution. In the next chapter, we take a closer look at the impacts of WFM on the people in your organisation.

---

1 Google Patents, Bundy Clock, viewed 31 July 2017, <https://www.google.com/patents/US452894>.

2 Schwab, K. 2016, *The Fourth Industrial Revolution*, Penguin Books Ltd, London, UK.

3 Ventana Research, op cit.

Chapter 2

WHAT DOES WFM MEAN FOR YOUR PEOPLE?

In business, it's tempting to look for some sort of technology solution to solve an intractable problem. In fact, every day software vendors the world over spruik the latest and greatest technology 'solution', desperately seeking customers to use it to solve their problem.

The drawback with this approach is, if a business constantly looks to technology to solve every issue, it overlooks a critical component of any successful business: its people. The key to solving most business problems (and I acknowledge this is a generalisation) is the need not just for processes, but people; not just logic, but emotion; not just technology, but humanity.

Solving business problems

So, what are the *main* WFM problems that businesses encounter; and what benefits can a company expect to accrue if it can solve them? To break it down, let's start by looking at the issues through a 'business first' lens and explore how WFM-related issues rarely occur in isolation, but cut across a number of internal and external business functions.

There are four main business problems that WFM solves (see Figure 2.1). At the broadest level, WFM enables business to solve each of these issues by:

→ obtaining deeper visibility of the workforce;

→ improving experience from both an employee and customer perspective;

→ creating more flexible work opportunities; and

→ ensuring compliance of pay and entitlements to reduce risk.

Each of the problems described in Figure 2.1 are generic in nature, but they could be aligned specifically to your business. Whenever problems need to be solved, they should be solved looking through the eyes of the industry or the profession where they exist. I call this 'business first solutioning'.

Figure 2.1: The four main problems WFM can solve

	Visibility	Experience	Flexibility	Compliance
Challenge	You lack visibility or confidence regarding the profitability of the aspects of your business.	You're not sure if you are losing sales or your customers are waiting too long.	Your staff demand greater work flexibility or you risk losing them.	You have concerns over the accuracy of staff pay and entitlements.
Benefit	Complete reporting and analysis of costs to forecast and make real time operational informed decisions.	Increased accuracy of forecasting and planning to ensure team members are available to meet customer demand.	Provide staff with more flexible working arrangements and personalisation to develop and grow in your business.	Deliver accurate compliance for the correct implementation of awards and conditions.

Let's look at some real-life business examples across different industries to see how these problems manifest and how WFM can effectively solve them.

Service-based business

Challenge: Jennifer runs a service-based business. Her business runs on tight margins and she often finds that on completion of projects for her clients, unforeseen labour costs impact profitability.

WFM solution: Jennifer requires a complete picture of her workforce and needs to understand where time is being spent, right down to where it is being spent on each activity.

Benefits:

- → Jennifer's operations team can compare job forecasts to actual outcomes and make daily informed decisions to ensure only those team members required to complete the work are used, minimising the cost of labour.
- → Jennifer's operations team can also assess the profitability of each project and make quick adjustments to the forecast if certain activities are over or under budget.
- → Most importantly, Jennifer and her senior management team can bid for new jobs confident that they will be maximising profitability.

Retailer

Challenge: Graham owns several retail outlets, but he doesn't have enough retail assistants in his team to deliver the level of service his customers expect and deserve. As a result, prospective customers are walking out, frustrated, and Graham's existing loyal customers are starting to shop elsewhere. To add to Graham's headaches, his retail assistants are overworked and frustrated they can't keep all the customers happy.

WFM solution: To help solve these problems, Graham first needs to accurately capture sales volumes, foot traffic and other key metrics in each store.

Benefits: These metrics can then be used to create a forecast to align staffing requirements, maximise sales and minimise labour costs as well as provide customer and team member satisfaction.

It sounds simple when you put it like that, but these issues present huge logistical problems for retailers like Graham the world over and demonstrate the immediate positive impact that WFM can have in a retail environment.

Healthcare provider

Challenge: Dr Clarke runs an aged care facility. Nurses are overworked and angry that they don't have enough flexibility in their working arrangements. Their frustration has led to an increase in sick leave, an increase in workplace culture issues and has the potential to impact patient care if left unchecked.

WFM solution: To get to the heart of this issue, a well-organised WFM process would allow Dr Clarke's nurses to request their specific shift preferences; these preferences could be taken into consideration when rosters are planned.

Benefits:

→ This WFM solution balances the needs of Dr Clarke's healthcare facility along with the needs of her nurses and their patients.

→ Additionally, when rosters are communicated to nurses, an optimal system would allow Dr Clarke's nursing team to accept or reject shifts and quickly find substitutes if required. This approach would lead to an improved work–life balance for nurses, resulting in happier nurses who positively contribute to, and grow within, Dr Clarke's health organisation.

Let's look now at how WFM plays out in your organisation.

What WFM roles do your people play?

Solving intractable workforce-related problems requires people, process and technology skills. The list below (though not exhaustive) gives some sense of the multiple disciplinary skillsets required at every level of an organisation to drive the best possible WFM outcomes.

Looking through a business lens, this might cascade in the following way:

→ **Senior management** identifies a business problem that may be fully or partially solved by the use of WFM. It could be used to reduce costs, increase revenue, or maximise time spent with customers. This is often voiced to the organisation through a business strategy or an executive-sponsored project.

→ **Operations managers** are tasked with addressing the problem and implementing a WFM solution. They need the skills and support to manage up and down in the organisation, and the ability to recruit and develop the right talent to deliver the outcome.

→ **Team members** in the organisation need (a) skills to support delivery of the solution to the problem; and (b) to be in a work environment that gives them scope to learn and grow.

→ **HR** supports the organisation's talent to ensure the right people are recruited and compensation is aligned, organisational and departmental goals are aligned and appropriate compliance measures are in place.

→ **Industrial relations** (IR) supports the organisation when decisions are required that impact the employees and their industrial instruments. This role is often seen in organisations where unions are active.

→ **Payroll** provides the services to ensure legislative compliance and financial integrity and ensures people are paid in a timely and reliable fashion.

→ **Finance** provides the required financial metrics to report and measure the success or failure of areas of business or WFM initiatives.

→ **IT** provides the systems and technologies to support WFM or other business solutions in a cost-effective and flexible way.

→ **Business specialists**, either internal or external, are often required to bring all these components together harmoniously to create a working collective, focused on delivering a shared outcome.

Organisational impact during a WFM initiation

Taking these descriptions into consideration, let's look at how WFM decisions play out at each of the 'people levels' within an organisation.

Senior management

Typically, senior managers have a shareholder mandate to maximise revenue, minimise cost, reduce risk and use their available talent wisely, with a focus on quality and broader social considerations. At the senior management level, reference to a project is usually the easiest way to show the direct benefits of implementing WFM into an organisation. Senior managers approve the project, release the funds and empower a project team to deliver a defined set of benefits.

Once WFM initiatives have been implemented, senior management can also benefit from WFM to help with ongoing business decisions. For example, if part of a business is non-profitable but its people are working productively and efficiently, this may lead to a decision to re-purpose that part of the business.

Operational managers

The responsibility then passes to an operational manager whose job it is to define the detailed requirements to deliver the project. Benefits accrue if the operational manager can effectively manage their talent to achieve productivity improvements across common processes and functions. This should result in improved employee satisfaction through better staff engagement, stemming from optimised processes and functions.

Team members

People involved in the WFM process need to understand the reasons for change and how by adopting the new processes and functions they will gain greater visibility and flexibility in their working life.

This all sounds simple but, in reality, it rarely works like this. In my experience, misalignment can often occur because senior management tends to focus on the benefits contained in the project and then release the funds to move forward. At this point, senior management might believe their job is done and play no further part in implementing the process aside from making the occasional appearance at a steering committee meeting.

Let's use a quick example to unpack this.

An international manufacturer has employed a team of delivery consultants to start designing a WFM solution to address specific time-management objectives, which senior management has identified as a key business outcome required from the project.

Jim, an operational manager, doesn't understand why he has been invited to a workshop on time management. In the past, administrators and payroll people have always made decisions on time management. It appears to have no relevance to him.

The project benefits stem from standardising the time and award interpretation across different geographies, coupled with implementing standardised processes and functions for operational management to manage their staff. Senior management have not told Jim this, nor are the consultants aware of it.

Additionally, the award interpretation has been done the same way for years and is not strictly in line with the award. Further, there are new functions such as 'rounding' that the WFM system requires, but which also raise more questions for Jim: Why am I here? Why do I need to answer such questions?

The consultants are looking for firm answers to questions such as:

→ Is the remuneration the company pays its employees in line with current geographical practice?
→ What rounding rules apply if someone arrives before or after a scheduled start or finish time?

There is no clear understanding of who the decision-maker is; is it Jim, HR, payroll, IR or finance? Meanwhile, the project team are unable to agree and/or make a decision.

These questions are generally highlighted as risks, then they become issues, and often escalate to the governance layer of the project. Meanwhile, Jim and the broader operations team who he represents are confused and the consultants are frustrated that there are no clear answers coming from the manufacturing company. At the governance layer, the business project sponsor – who is still not 100 per cent clear about what their role is – allows current practice to remain in place and doesn't deal with Jim's concerns, as he is seen to be disruptive.

I think you can already see how a project can quickly enter a confused state, and the likelihood of deriving the business benefit is lessened. The key point here is that setting up a correct project structure with clear understanding of what business benefits you are looking to achieve, along with what you expect from your people, are important criteria for success.

Delivering the outcomes

Once a business commits to delivering improvements via WFM, it embarks on a transformation program. As I mentioned in Chapter 1, a key aspect of large-scale change and delivering stated business objectives is adoption. Key performance indicators (KPIs) or goals are valuable to ensure the overall business benefit is achieved, ideally via changes to job roles, measures and incentives. Senior management should be continuously involved in the implementation, to ensure buy-in across multiple areas of the business. To implement KPIs or goals, the broader business areas and systems may need to be realigned to enable the changes and deliver the outcomes. This will help ensure the benefits flow to the operational manager and staff, who will in turn better understand the benefits being delivered.

To deliver true transformation is not easy and should not be underestimated. It requires senior management commitment and leadership to achieve these outcomes.

Top take outs

- → Look at all problems through a 'business first' lens and provide a 'business first' solution.
- → The four key benefits from WFM are greater visibility, improved experience, greater flexibility and improved compliance.
- → WFM cuts across numerous business functions and roles, which are all required to work collaboratively to deliver an outcome.
- → Industry knowledge is a major consideration in any WFM initiative.
- → Clarity on project outcomes and people expectations will help your WFM project get off on the right foot.
- → KPIs or goals will enable you to streamline the way business benefits are achieved for your organisation.

Where to next?

This chapter has provided an understanding of what WFM means for your people if you implement it properly – and problems that can occur if it is not. Chapter 3 looks at the application of these benefits from the perspective of your customers.

Chapter 3

WHAT DOES WFM MEAN FOR YOUR CUSTOMERS?

Over the years, I've often met with prospective clients who want to hear about and learn from my previous experiences implementing WFM processes. So when compiling this section, I thought a client case study would be a great way to help readers understand the types of benefits they can achieve for their customers as well as their people. While each industry is unique, as are the outcomes each business is trying to achieve, WFM provides a common set of underlying benefits. You will recall I introduced these in the previous chapter:

- → Greater VISIBILITY
- → Improved EXPERIENCE
- → Greater FLEXIBILITY
- → Improved COMPLIANCE.

In short, the benefits of WFM are far-reaching and have wide organisational impact. Let's look at some of these and their flow-on effects that result in improved customer experience.

Far-reaching benefits

Providing improved processes to people enables improved efficiency. These efficiency improvements allow time to be channelled away from administrative tasks towards value-adding tasks. Efficiency improvements make the workplace more enjoyable for your employees, which enables them to become more inspired about what they do, and this is reflected in their attitude towards your customers.

Your management team has a responsibility to clearly communicate its organisational vision and purpose to employees; this will inspire them and enhance their pride in your organisation's brand, which they convey to your customers.

As you can see, once your business has an inspired workforce, it can improve customer experience as your people believe in what they do and why they do it. This permutates to strong engagement with your customers and will result in a highly personalised customer experience. Personalised experience results in improved customer loyalty to your organisation's brand.

WFM allows you to track external factors and use this data to improve customer experience. For example, if you're in retail this may include omni-channel data on previous buying trends, the customer's behaviour in the store, their likes from social media collated in real time and sent to the store assistant as part of the personalisation experience.

Let's examine this further with a look at a case study of high-profile retailer, T2 Tea. T2 has completed a WFM digital transformation and in the following interview with Georgegina Poulos, Global Director People, T2 Tea, she explains how T2 was looking to WFM to:

- improve experience for their largely casual workforce;
- collect accurate data to make better decisions and remove time-wasting administrative activities; and
- ultimately, ensure that this all resulted in their people creating an optimal experience for T2 customers.

Interview: The Customer – Georgegina Poulos, Global Director People, T2 Tea

About T2

T2 is about reinventing and reimagining the humble tea leaf, and sharing our teas with anyone who'll listen. We're inspired by the people we meet and the far-flung places we visit. We get a kick out of taking ancient tea rituals and reimagining them, bringing them to a modern tea table. Every cup we brew is a chance to make tea more enjoyable, more accessible and more experimental; it's our opportunity to connect, understand and share with the world our love for a better cup of tea, every day.

T2 has 90+ stores globally, including in Australia, New Zealand, the United Kingdom, the United States and Singapore.

About Georgegina Poulos – Global Director People

Georgegina Poulos is a senior executive with a strong human resources focus and extensive experience in both the international and domestic arenas across a variety of industries. Georgegina's expertise is in organisational development, training and development, leadership, culture-building and change management. Her key strength is that she combines excellent organisational and communication skills with a strategic business approach to human resources. As an organisational disruptor, she has a proven track record in assisting organisations achieve sustainable business growth and a positive operating culture within complex multi-site environments.

Georgegina holds a Masters in Applied Sciences (Organisational Dynamics) and a Postgraduate Diploma in Human Resources and Industrial Relations.

Jarrod: Thank you, Georgi, for speaking to me about your workforce initiatives. In opening, what does the term workforce management mean to you?

Georgi: For me, it's really about being strategic in understanding who you need to have in the right role at the right time. Then I think you can break it down into what's the current need versus the future need, and then where are the gaps that you need to build for. I think more in a holistic way rather than a one-by-one element of the HR process. Thinking about it in terms of really understanding, if we've got a business - and in this business, retail - what do we actually need to execute the strategy who do we need and when do we need them?

Jarrod: What prompted T2 to embark on a program of work in this area in the first place?

Georgi: We're a global company - we're now in five markets - but we don't have great data. We're a founder-led organisation, established 21 years ago. We've grown over the last four years as we were acquired by Unilever, and we've almost doubled in size since 2013. That's put a lot of pressure on the resources, but we've done it. And we've sort of just said, OK, well this is how we've always done it, we just need to continue that pattern. It's not sustainable, so we're in a paradox now of having this tension of wanting to continue to grow, but also needing some governance.

We had a need, and the need was really around payroll. As we started to unpack payroll, we started to think actually, it's time and attendance data that we need.

We don't have a central source of truth. We have an e-recruit system; we have a learning management system; we have a time and attendance system in ANZ; we have manual processes. I think we've got nine different touchpoints, different systems in the people's space, and none of the data talks to each other. I can't realise the people strategy or the organisational strategy if I don't have the right data to help my leaders make sound business decisions about their people. We had to start somewhere, so we started because the need was greater in the time and attendance space, with the retail network being a larger portion.

We've decided that we'll start there, but we will go back to that holistic picture. It's an education process in this business. We're not tech-savvy in a lot of areas. When you're not a revenue generator in any business, you're a cost centre, how do you actually build a business case for investment so you can prove your worth to get the cost benefit at the other side? You have to be commercially savvy to sell that business case. For me, it's all about being able to support the business in being more commercial and making better decisions, and having the right data so my team can be true business partners and not processing-input team members.

Jarrod: I remember in one of our early discussions, you said that you were looking to make the leaders in stores more commercially savvy.

Georgi: Absolutely, they need to be. Entrepreneurial spirit is something that we harness at T2. We're a founder-led organisation; we like to take risks, we're a little on the quirky side, and we do that really well. We move really fast - changing gears is normal for us - but we really focus on that experience that the customer has, and that the team has. We need to shift the dial and continue to keep the customer first, and our team members and leaders need to savvy up in that commercial space, and give them the tools so they can be entrepreneurial. If they're not understanding their budget costs or their labour costs or their loss-prevention costs, then they can't really be entrepreneurial. Technology can really be helpful here. We can produce the data for them [the stores] so they can actually run their business by being business owners, learning new skills; making sure that we've got the right people on at the right time to make sure the customer gets it, so it's all about the customer. It's not about cost-cutting, it's not about number-crunching, it's about giving our leaders access to data and helping them make decisions that benefit the customer.

Jarrod: Can you tell me a little about the T2 business and specifically the culture?

Georgi: It's a great culture. We have five fantastic values that we bring to life in as many ways as we can. We reward those values, we live them. It's easy to say somebody's not T2, than to articulate what T2 is, like with any culture, because there's just something magical about this place. It is because we're pushing boundaries, we do things differently to celebrate difference and diversity. And our people make that difference. Our product is fantastic; we have a very established market here in Australia and we're growing globally, but it's the people bringing that experience to life, so making sure we are recruiting, developing and retaining the best talent is paramount. Understanding what's in the market; what does looking good look like; how can we develop our people; how do we retain them? This is all paramount in giving that customer the ultimate experience.

If we start to look at team members around 16, 17 years old, we're creating a customer for life, because we're starting to consider our mission of creating a T2 generation on every continent. If we can get them loving tea at that age, even if they don't work for us, they're going to be connected to us. It's twofold. I think Nokia did it in the States many years ago. They actually gave out mobile phones to high school students, and they created customers for life, because it was new technology at the time, so I'm talking a little while ago. It was also about finding out how to connect them to a brand. If it was their first phone, they most likely stayed with that brand because of the emotional connection. You must think of your team as your customers.

Jarrod: How does technology and the rise of digital influence T2's WFM strategies and the business?

Georgi: Globally, our workforce are Millennials. We have an extremely young workforce. They've grown up with an iPhone, a laptop, a tablet. Our team members are tech-savvy. We need to be able to adapt our processes and our mediums that support this new workforce. This new workforce expects things to be digital, expects it to be any device, anytime, anywhere.

We can't live in a world where we're expecting people to do manual leave forms or manual change of status or sending in resumes. A, it's not efficient for us. B, we're not capturing data. It's not right for the environment because it's too much paper. Our consumer doesn't live that way, our team member doesn't live that way. I think about, what do I get that's paper these days? I get all my bills electronically, I pay everything online, I order food online and have it delivered. I don't really do much with paper unless it's a legal document, and even then you can scan your signature. We've had to adapt our processes so that we're saying, you're no worse off. If you can do this at home, you can do this in the workplace. We have to use digital technology, and it has to be mobile.

Jarrod: Has your Millennial workforce pushed you down this path, or are you driving them?

Georgi: Great question. I would like to say that though I'm not a Millennial, I'm pretty tech-savvy. I use social media and digital platforms a lot in my own life. The first time I implemented a human capital system was in 2006 – a little bit ahead of the curve for some businesses. I've been in the space a long time to understand the benefits of it. I think in some respects this workforce is driving us, because it just makes the adaptability to change so much easier. There's less resistance, therefore you can take more risk. When I was implementing that human capital system, 70 per cent of my workforce was blue collar. They were computer illiterate, and actually, they didn't work behind a computer. We had to set up kiosks, we had to do a lot of development in order to get change. It was really quite aggressive, if you like. We were taking a workforce that drove trucks; however, we were going to be sitting at a computer. I'm like, I work a forklift, I work in a warehouse, whereas our workforce all have technology.

I think that they're leading us, but we're also saying, 'you know what, we can do this now', because the change of adapting to a device isn't the challenge. It's adapting some of the processes and the governance around that and making sure it's user-friendly, the language is right, and it can be adapted across multiple markets. It's a bit of push and pull. Our team's not stuck in the dark ages where, 'I don't want to do that'. For me, it's around the big picture of having everything connected, but also managing my own passion for change to when the workforce is ready for it.

Jarrod: Just thinking back to that 2006 implementation you mentioned, do you recall how long that took?

Georgi: Yes, I worked seven days a week on it for 11 months.

Jarrod: And this project by comparison?

Georgi: Wow, so this project, end-to-end with the Australia/New Zealand/Singapore rollout, will be three months. Yeah, three months!

Jarrod: That's where the cloud has improved things and enabled this rapid transition.

Georgi: You just can't operate, I think, as a successful modern contemporary agile business if you don't have technology. It can't just be customer-facing. A lot of businesses, say, have a one-customer view. I adapt that and it's one system, one team member. One team member, one system. You have to have a one-team member view, because there are so many stakeholders that need data; whereas to input into other systems, other processes, reporting, whether it's at T2 or Unilever, we need to be able to understand what our workforce is like, and what risks are associated with that – particularly when we talk about talent. This year I've introduced a talent management framework, and won't it be awesome, and it will be one day; it will be online so that we can actually have a just-in-time moment of our pipeline.

You know, the minute you put data into the spreadsheet, it's dead. That's where we are at this moment in time, because talent management and succession planning is new for our leaders, so we need to educate them around the process. At least globally, if I had the system, we would be doing it online and quicker.

Jarrod: What is the biggest issue you've faced in terms of managing your workforce effectively?

Georgi: It's not technology, actually. It's more about society; specifically, the increase of mental health issues. My generation of workers, yes, we would have experienced mental health issues, but really didn't talk about it. I now have to really try to unpack, 'I've got anxiety' or 'I've got depression', and find out what's the underlying factors to that. How do we equip our leaders and team members to be resilient in a world that's ever-changing? Part of it is – and I'm not saying I don't love technology – but you're always on. How do you actually turn off? Because if you have any device, anytime, anywhere, you need to build in practices to give yourself time to relax and switch off. I think that's a real barrier not just for us, but in society. There is the increase in mental health issues, due to the pressures that are put on people because they are so much more accessible and don't have time to switch off.

We don't have any problem attracting talent, so we are lucky. We are a brand people want to work for, and it's about making sure that we're getting the right people at the right time and that we need. [We are] super lucky in that respect; there's not a shortage of talent. Retail in

Australia isn't seen as a long-term career like it is in the States or in Europe. We do have some element of a transient workforce. As people are studying we may have some transient workers, like in hospitality where you might only get them for a year. We've got some people who have worked with us a good amount of time, like three or four years while they're studying. I still think it's great, it's not that transient. You do have waves of people, students [for whom] it's their part-time role. Some stay on and move into leadership roles, and we have a really strong pipeline internally where we do promote within where we can. It's about continuing that pipeline, looking at our mix of workforce and, with an 80 per cent casual workforce, actually asking 'is that the right mix to be building succession'? If we had more permanent or part-time roles with the opportunity to develop leadership and managerial skills, what would that pipeline look like? So I think that's something we need to consider around managing that.

How do we take this talented workforce that might be studying areas that we need at HQ, and how do we use that as our graduate pipeline to bring them in? That's all a work-in-progress for us. We just introduced an internship program; we'll be looking next year at what a graduate program might look like. We advertise roles internally. We are based in Melbourne and we have teams across the globe. We need to be thinking about where are the hubs, where could you be and work remote? [We need to be] shifting the focus of being in the Melbourne office. Some roles add a real value that are based here, so it's a real balance. As much as we want to make sure everyone has the opportunity, you do need to balance that with what are the needs of the business; what are the strategies; what do we need to deliver; and then how do we deliver that?

Jarrod: What is the biggest opportunity in relation to workforce management that you see now or in the future?

Georgi: I think really understanding our people. You've employed people for a role, they come in, [so we need to use] a human capital system that can capture aspirations, career aspirations, hobbies, external curriculum that adds value, so that we're really getting to know our people beyond how we're just recruiting them. [We need to] have it in a place where we can actually draw upon it and give the team member the opportunity to almost have an online internal resume, where they can opt in for projects or use skills they might be learning at school in a different way. I think it's about real opportunity, really knowing our people, and seeing beyond the role they're in.

Jarrod: It's really making that experience personalised?

Georgi: One hundred per cent, because they might have been a firefighter in a previous role and we might have an emergency management team. That might be right up their alley, but unless we ask those questions, we never know. We've got this rich, deep, educated workforce that we can tap into a whole lot more locally.

Jarrod: Do you see training and education going that same way in a very personalised fashion as well?

Georgi: I do. At the moment, we do a lot in the retail space. We do a lot of education around product and customer experience and really engaging with our customers. The focus for 2018 will be a lot more around leadership and commercial skills. We've identified that as a gap. We also want to equip our leaders with the right legislative knowledge, whether it's around recruitment or employee relations or IR or whatever it might be. We do want to have programs that are allocated, but we also want to have sweet programs people can opt into. We're in the process of rolling out our competency framework, where we have developed 26 job families, and each of those 26 families will have a success profile attached to it. A success profile is made of four quadrants. What experiences are required, what are the competencies required, what's the motivational effect, and then what organisational knowledge you will need so you can then map those competencies to programs.

If somebody is in retail, they might be a team member, but they'd like to be an assistant manager, or they're an assistant manager and would like to be a store manager; you can really look at those profiles, and if we have programs mapped to the competencies, they can actually manage their own career pathway. It could be personalised, or it could be in the group setting.

Jarrod: If another business was about to adopt a workforce management transformation project, what advice would you give them?

Georgi: Really for this piece, it's got to have a brand. It needs to have its own identity so you can really have some fun with it to gain traction. We're going to be calling ours 'Belong2', because you belong, we belong, we can really tell a story around that. I think you need to understand your imperative for change. What is it that is really in it for them? What are the factors that are getting you to change, and what does success look like? Leaders who are leading the change need to be storytellers, bringing the team along, because change is scary for anyone. If I know why I'm doing it, what's in it for me, what I'm going to get out of it, and what role I need to play, then I'm more likely to get on board. Do your homework, see what system suits you and your business. Don't be seduced by the big players, because there are some other systems technologies around that might be more suitable for your business. Really get your stakeholders involved in the infancy of the project.

Jarrod: Would you also see an importance of putting a KPI-framework in place that cascades through your organisation from senior management level, to help with the alignment?

Georgi: I think that old saying 'if you can't measure it, you can't manage it' has merit. Don't get sucked in by just 'we've got so many people using it', because they might just be using it to tick and flick. They've got to be really meaningful KPIs that are actually aligned to strategy.

Jarrod: Where do you see the future for workforce management, blue sky?

Georgi: The notion of portfolio worker will creep into Australia more and more. And what I mean by that is really taking that freelance model; so instead of working for yourself and working for multiple companies with an ABN or a contractor model, actually being an employee of multiple companies and doing the work that you love. That's where I see myself.

Jarrod: I've heard of a Personal Data Repository in the past. The concept is that you are essentially responsible for your own people master data, and so your history of what you do and all your competencies follows you, and you can make that available to other organisations through your workforce.

Georgi: Yes, I think LinkedIn does a little bit of that, and I use LinkedIn like that. I actually host all my data pretty much there and then adapt my resumes and stuff. I think that's exactly right. You need to own what you bring, and that notion of 'portfolio worker' is that you're never going to be 100 per cent satisfied with any job. There's always going to be some stuff that you don't really love doing, but you need to do it, because it's part of the role. I think, more and more, we will have higher engagement, and not every role will be able to manage this, but I think we have more opportunity to tap into people on more flexible terms – to keep them engaged in the organisation, but only look at what they really love to do. For me, it's strategy. I could definitely do people strategy in four or five different organisations, and that's where I see the future of work.

Jarrod: One final question. I think in the past I've seen, very much from a business point of view, that people look at things from a functional perspective; like 'I'm payroll' or 'I'm HR' or 'I'm workforce' or 'I'm finance'. What do you see the future becoming? What I'm really getting at is, do you see it becoming much more process- and outcome-centric as opposed to functional-centric?

Georgi: Good question. Part of our job families are more process or job role as opposed to title. For example, we've created a job family for a technical expert, for a project manager. You might have 10 or 15 people in that job family that do really different things functionally or are called something different, but the drivers or the types of activities they do are very similar, so therefore similar competencies are required, and then specifics are done in position descriptions or whatever. I think it has to be a balance because you don't want to work in isolation and go, 'this is all that I do'. In some cases, you need technical expertise, and that's what I bring. It's about getting people to excel at the level that they're at, and always wanting to challenge and try things differently without trying to be this massive generalist.

Jarrod: Thank you so much, Georgi.

* * * * * *

Top take outs

- WFM is strategic in its nature and enables business strategy to be delivered through business operations.
- WFM initiatives can provide tools for managers and leaders to become more commercially savvy.
- WFM enables store leaders to make decisions that benefit the customer.
- A digital footprint is at the core of business initiatives as it's at the heart of how employees and customers interact.
- Be cognisant of your team, the customer and their needs.
- Business benefits can be realised quickly, provided the business is ready to change.
- Lead WFM initiatives with a brand to enable the story of benefits and 'what's in it for me' to be told.

Where to next?

This chapter explained how WFM plays out from a customer-adoption perspective. We looked at this through the eyes of T2 to highlight the benefits you can achieve from WFM. The next chapter looks at the world of AI and what you might need to consider in its application.

Chapter 4

THE ROLE OF DIGITAL AND ARTIFICIAL INTELLIGENCE IN WFM

Much has been written in recent times about artificial intelligence (AI) and what it means for all of us. Is it a real possibility that machines will replace humans? A more realistic scenario is that machines will help us become more efficient, make our lives easier and be used to solve specific business problems.

In this chapter, I want to examine AI and the part it is playing in WFM. I gained some insight into this during an interview from world-leading AI futurist, Matthew Michalewicz, who put some of our thoughts and fears about AI in context. Matt's final comments regarding the capitalist society that is driving the charge in AI are perhaps the most thought-provoking.

But let's start this chapter by asking:

What is artificial intelligence?

You will find many definitions of AI but, to me, AI in the context of WFM is a collective term that incorporates the areas of robotic process automation (RPA), machine learning and deep learning to automate logical tasks that generally require human intelligence to solve a problem in areas such as visual, hearing or speech. Let's look at a simple application of AI in the workforce to introduce this topic.

5-step manual process to request a shift swap

1. Employee contacts manager and requests to change their shift
2. Manager checks for a suitable replacement with the same skills and contacts them to check availability
3. Replacement employee confirms availability for work
4. Shift swap is enacted
5. Employee requesting the swap is notified of approval.

Applying AI to this process removes the human intervention by the manager. Here's the revised, automated process.

5-steps process to request a shift swap using AI

1. Employee requests a swap for their shift via an app on their phone
2. Technology finds a suitable replacement with the appropriate skills who is available for work and contacts them electronically to check availability
3. Technology confirms the shift swap with replacement employee
4. Technology enacts the shift swap
5. Technology confirms the shift swap with the requesting employee.

In the AI process, there is only one human interaction at the outset of the shift swap, and there are numerous benefits:

1. AI removes the need for human back and forth to check and confirm the change
2. This frees up the manager's and employee's time for value-adding activities
3. The process ensures compliance and audit
4. It cuts down the time the process takes
5. There is a digital trail and data science can be applied to identify efficiency improvements into the future.

Workforce robots

I've mentioned already that we will see the introduction of what I call 'workforce robots'. These workforce robots will automate jobs that are currently undertaken by a human. For example, a workforce robot may take the front-line call in a contact centre and engage in a conversation with the caller to determine where to intelligently route the call.

These workforce robots introduce a human dynamic to the workforce. If they break down, analogous to a person being sick, there will need to be a suitable workforce robot replacement. Alternatively, if there is no workforce robot replacement, a human replacement will need to be found to avoid interruption to the process. The major implication is that customers will no longer be able to interact with an organisation if there is no 'worker' (robot or human) there.

Workforce robots will need to be scheduled to complete a task, just like a human. This will be necessary so the business knows that the role associated with the front-line call is being fulfilled, even if it's by a workforce robot.

You might think workforce robots are akin to the robots that replaced workers on production lines years ago. But there's a key difference. Those robots *replaced* humans on the production line - these jobs were automated, never to be completed by humans again - the role became part of the physical machinery used for production planning. Workforce robots, however, remain part of the human-based, planned workforce. They are interchangeable with human workers, rather than a replacement for them.

Robotic process automation

Another feature of AI is RPA, which takes tasks that are completed by people and automates them. For example, think data entry associated with time entries; manipulation of data; communicating digitally with other systems and so on. These are processes that can be taught by a human using software and executed by a machine.

RPA can deliver many benefits, including:

→ increasing quality of work as it eliminates human errors;

→ improving productivity as the process is always on; and

→ reducing costs by removing many human components of the task(s).

Machine and deep learning

Machine learning builds on the concepts already in place for AI and uses mathematical and statistical methods to find hidden insights and make predictions. For example, if an employee is continually late for work every Monday, a performance management record may be triggered for that employee's manager to speak to the employee to take remedial action.

Deep learning builds on AI and machine learning to *continually* learn, just as the human brain does.

For example, an employee might habitually request to work each Wednesday afternoon. Using machine-learning techniques, the system will be able to predict this request and make the process more efficient by automating this request each week.

Over time, the deep-learning techniques may identify that while working on Wednesdays, the employee is 20 per cent more productive, and the deep-learning algorithm determines that on Wednesday they complete their allocated work in a shorter shift, resulting in the same quality output. This learning is then applied to optimise shift lengths for all employees, resulting in decreased shift length, lower labour costs and the same quality output.

The benefits of machine learning and deep learning are far-reaching. For example, workforce robots will learn from human interactions and develop their knowledge base, and predictive reporting will offer cause, effect and solution options to people.

* * * * * *

To understand the part AI is playing in WFM, we next have to consider:

What is digital WFM?

We are seeing technology advance at such a rapid rate that organisations and people are struggling to keep up with it. The internet of things (IoT), where we see networks of connected devices, bitcoin, financial transactions without the need for a third party, driverless cars and masses of data are becoming part of everyday life.

Into the future, nearly everything we do will have a digital footprint. Any transactions we undertake will almost certainly be driven from a mobile device pulling all sorts of omni-channel information networked from anywhere around the globe or in space, helping us make informed decisions.

Let me paint a picture of what this might look like from a WFM perspective, with a description of an end-to-end process.

1. Imagine you and your team are working in a service industry and your next outdoor job – fixing a leaky roof – has just been automatically rescheduled because rain is forecast within the hour. It's important your team fix the roof prior to the rainfall.
2. You are instructed to head to the service depot to collect some parts, as your van stock is missing materials required for the repair.
3. You drive into a loading bay and a workforce robot loads the materials into your van.
4. While you are there, a sensor has detected that your tyres need air, so the tyres are refilled.
5. When you leave the depot, the job tracking system knows who is in the van with you due to wearable technology.
6. Once you arrive at the job, the start time automatically gets recorded because you have automatically triggered a job start at the defined GPS location.

7. One of your team members goes home sick and a replacement team member with the right skillset is automatically diverted to your job from the closest location.
8. The materials you use are automatically tracked against a defined bill of materials for the work.
9. Once your job is complete, you drive off, automatically recording the time as you depart.
10. When you are driving to the next job, you dictate your completion of job report, which is automatically transcribed. You also dictate your job notes, record the materials you used and note any occupational health and safety (OH&S) issues.
11. This information is collected centrally to ensure you and your team are paid correctly for completing the work, and any replacement materials to replenish the van stock are ordered.
12. This job information is then statistically analysed to determine how long the job took, taking into consideration the weather and your fatigue rating.
13. The analysis looks at skills required for the work, time worked to complete each of the tasks and your overall health.
14. Adjustments are made to the standard work order and time required to complete tasks automatically based on your actual data, including team utilisation.
15. OH&S recommendations are automatically logged.

You can see that in this process AI plays an integral part in digital WFM. For example, tasks are set automatically and recorded digitally, workforce robots assist in loading and unloading parts necessary to complete the job, machine learning techniques are used to analyse the job data and optimise the standards in place for the time taken to complete jobs.

Ultimately, what we do and how we do it is becoming smarter. Learning will not only be by humans, but by algorithms programmed by humans to analyse work patterns and make work output more productive in real time.

Digital WFM – changing the way we work

I'm hoping that by now you are starting to understand the potential that digital WFM, supported by advances in AI, has to change the way we work. The changes will continue in greater numbers and at a faster rate as they are embraced in our workplaces and we start to value the benefits that they bring.

Let's look at some of these changes.

Networks

The world is now essentially one large, interconnected network that will continue to grow. People are connected, machines are connected, information is connected. We have moved from a world of person-to-person connection, to node-to-node connection.

These changes bring many benefits: we have access to a world of knowledge; we can use this knowledge to learn; we can mobilise quickly; we can flex up and down quickly; we can see things in real time, and so on.

This means the way we work in the future will become highly responsive, focused, agile and available 24/7.

Remote work

The network concept also enables us to mobilise highly specialised people with specific skills across the globe, in real time. For example, Ellen has a requirement for a retail specialist who understands radio frequency identification (RFID) technology. She needs to understand how this technology interacts with shelf-fulfilment algorithms and auto-task generation activities in her store. Ellen will be able to use digital technology and seek out someone with these skills anywhere in the world, efficiently and effectively.

Remote work will become the normal way of operating. Collaboration technology has reached a tipping point where you can virtually interact as if another person were in the room with you. Augmented reality and virtual reality will also enhance this experience over time. For example, the specialist will be able to see your store in 3D through a headset and communicate with you as if they were there with you.

Team-based work

Just as we are able to source appropriately skilled people from across the globe, the concept will extend to teams. Highly skilled teams with the knowhow to solve specific business problems will work remotely or on-site to complete a defined outcome. Picking up on the previous example of RFID technology, there could be a team deployed to implement this throughout your retail stores.

Self-management

Over time, the contingent, freelance and 'gig-economy' workforce will grow with people having multiple jobs, especially Generation Z and beyond – our future workforce. Generation Z appreciates life, is comfortable with technology, is ready to do things now, likes to be in control and can multi-task with ease. With this in mind, it will become an increased and desirable requirement of the individual to manage their own time and for employers to respect this. The individual will control their availability, and the organisations they work for will pick up the individual's availability from their own master data record.

One concept is a Personal Data Repository (PDR). A person's personal information, availability, skills, preferences, etc., are available to employers (and others they grant access to) via a platform. A realistic example of this might be: On Monday I drive for Uber; Tuesday and Wednesday I will be available to work shifts for retailers (remotely); Thursday I will be updating my first aid certificate, which I will store on my PDR; and Friday I'm unavailable (because I am competing in a surfing competition).

People impacts

To many, these changes are exciting, but others will be concerned about how they will impact the workers. Anyone considering adopting digital WFM in their workplace needs to be aware of the human impact and be able to explain to the workforce why these changes are being introduced.

Ability to adapt and adopt quickly

Quality technology is being developed at astounding speed. In turn, there is increased pressure on businesses to adopt technology and revise processes to remain competitive. This means that people must be flexible and quick to adapt to, and adopt, new technology and processes. Organisations need to be mindful of this and provide an environment that supports rapid adoption of new technology and processes.

Reskilling

In a recent post by the World Economic Forum[1] it noted that by 2030, 210 million people around the world (equivalent to the population of Brazil) are expected to change occupation and 800 million workers worldwide are at risk of labour disruption (see Figure 4.1). These massive changes are driven by forces including industrialisation, globalisation, digitisation and automation.

Figure 4.1: Skills disruption

Average

35% of core skills will change between 2015 and 2020

Disruption across countries and industries

%	Industry
43%	Financial services and investors
42%	Basic and infrastructure
39%	Mobility
35%	Information and communication technology
33%	Professional services
30%	Energy
30%	Consumer
29%	Health
27%	Media, entertainment and information

%	Country
48%	Italy
42%	India
41%	China
41%	Turkey
39%	South Africa
39%	Germany
38%	France
37%	Mexico
31%	Brazil
29%	United States
28%	United Kingdom
27%	Australia
25%	Japan
21%	Gulf Cooperation Council
19%	ASEAN

Average disruption (dashed line between 39%/37% and 35%/31%)

Source: The Future of Jobs report, World Economic Foum

We can prepare for this now by ensuring that we develop digital skills so that our teams are ready to adapt to the future of work. We also need to be cognisant that, with the introduction of automation, job roles will evolve and there will be a natural requirement to reskill our workforce. For example, in a retail store a robot can put away misplaced clothes while the associate concentrates on value-add, customer-facing tasks.

I'll finish off this section with a quote from the report that notes we are moving 'from coal miners to data miners'. For me, this really conveys the enormity of the transformation we are undertaking. Everything we do will be driven by data and data science.

Human resources

The digital changes that are all around us are reshaping HR departments. HR professionals have an amazing opportunity to lead these people-centred changes to provide strategic direction to their organisations. Business leaders will be looking for innovative ways to transform organisations and build a culture of continuous change, yet ensure people remain at the core of their business.

Education mindset

Moving forward, I see an entire shift in education. Traditionally, most of our theoretical skills were learned at university or centre of advanced education. Often, real-world skills were learned after university or at the centres of advanced education when we commenced work.

Today's generation is now googling its way through life, using lateral and logical ways to solve problems in an instant. If we work backwards, we now need to commence technology and business education at a very early age. Traditional curriculums will still have relevance, but new subjects and skills will need to be added.

Education will become highly focused to the task at hand, with subjects becoming personalised to the needs of each student.

Will there be less uptake of university degrees, as kids will have acquired a sufficient knowledge set to bypass this stage of pedagogy? I think so.

Interview: The Digital AI Futurist – Matt Michalewicz, CEO Complexica Pty Ltd

About Complexica

Complexica provides enterprise software applications that harness the power of artificial intelligence and big data to improve the effectiveness of sales and marketing activities, particularly for wholesalers, distributors and manufacturers, characterised by a large stock keeping unit (SKU) range and long tail of customers.

About Matthew Michalewicz – CEO at Complexica Pty Ltd

Matthew has more than 20 years' experience in starting and running high-growth tech companies, especially in the areas of predictive analytics and optimisation. He is currently the CEO of Complexica, a provider of artificial intelligence software for optimising sales and marketing activities, and a director of several ASX-listed companies, including Prophecy International (ASX: PRO), ComOps (ASX: COM), and LBT Innovations (ASX: LBT). He is also the author of several books, including Life in Half a Second, Winning Credibility, Puzzle-Based Learning, *and* Adaptive Business Intelligence Interview.

Jarrod: Thanks for your time today, Matt. Can you tell me a little bit about your career and how you got into the world of artificial intelligence?

Matt: I think the easiest way to answer how I got into it is that I was born into it. What I mean by that is both my parents are PhDs in math.
My father moved from math into computer science when I was six years old and specifically [into] a branch of computer science that's now known as artificial intelligence. That's creating artificial neural networks and machine learning, the whole idea of machine thinking and knowledge discovery and so on.

So I would finish school as a child and walk to the university, which was one kilometre away, and I would wait for my father to finish up. I played video games or drew on a piece of paper, while all of the conversations around me were AI, whether it was with PhD students, or lectures that he gave, which were discussions about the profession.

Fast forward 20 years, I graduated from the university and AI had come a long way, both in technological capability and also in understanding. My father had published, I don't know, 15 ... 20 books ... on the subject. All of those things converged. I also have a very close relationship with my father, where we created our first AI business together and commercialised what he had developed at the university.

That was a very logical ... it's kind of like your father being a fisherman and you became the fisherman, your parents are actors and you go into the movie industry, that kind of thing. But my parents were academics, whereas I'm not an academic, I'm a business person, but the field is one I've been around all my life.

Jarrod: You've experienced firsthand the evolution of AI; can you describe some of the key milestones you've seen along the way?

Matt: I'm not sure if it's a milestone, but the biggest change I've seen is the perception and what's called the 'understanding' of AI. I remember seeing presentations my father gave to business people who would come to the university in the '80s. I'll never forget when my father started talking about AI and one of the managers/executives from some company raised their hand and said, 'Excuse me, Professor. I just wanted to interrupt and ask a question. Is artificial intelligence ... like what is that? Are we talking about aliens here?' As funny as that sounds today, general understanding was so limited and so non-existent back then, even though there was a lot of research being done in the field; whereas today you've got movies on it, you've got Stephen Hawking talking about it, Elon Musk, AI in banking. They're replacing people; automation of jobs has become a mainstream topic. I think the biggest change in AI is the perception and awareness of it, and I think most people still don't understand what it is, but they're at least aware of it and they talk about it.

Behind the scenes, you've got a progression around the technology, especially in the area of machine learning, deep-learning algorithms that are coupled with computing power and the internet, which provides sources of digital data boosting text and its image to train machines on. All of that has allowed the progress technologically of AI. But that's not as great, mind you, as the market awareness and hype that's been created from when I was first exposed to it, compared to today.

Jarrod: What do you define as being AI?

Matt: You can break AI into two categories. General AI, which is what most people, like the non-technical person, when they think AI, they think *general* AI. General AI is still [unachieved] ... and not only is it unachieved today, but there's great disagreement among a whole lot of different people – not only computer scientists but biologists, psychologists, etc. – on whether it will even be achievable. General AI is the recreation of the human mind in a digital form that allows for emotion or emotional interpretation, decision-making, creativity, the ability to put that digital brain into a robotic structure, as that robot moves around, picks things up, and so on. Most people when they hear AI, certainly in the science fiction movies, etc., that's what they're portraying. In my view we are a long, long way away from that becoming a reality.

Then there's this other area of AI – let's call it *specific* AI – which is problem specific. It's the application of technology applying machine-learning techniques and so on, to specific problems. The problem could be detecting fraud in credit card transactions; it might be detecting patterns and images; it might be used in security. It's very specific in its application. I sat on the board of a company that has used machine-learning and AI to image and analyse those images and interpret them for culture samples in pathology labs. That's a very specific area of AI and the techniques are well advanced. They work, and they provide very, very quick results.

I think that's where a lot of misunderstanding or confusion arises. People see the movies, they see the hype, and they think of general AI; whereas the major advances today are really around the specific applications, like getting AI to beat someone in chess. These are the most common areas of application today.

Jarrod: That's a great differentiation. Thank you for that. When customers embark on a digital journey that includes AI, how should they go about it?

Matt: I was always taught growing up, you shouldn't look at it from the point of the technology, you should look at it from a business outcome perspective. In other words, if you take, for example, a retailer, the first question to ask – forget AI or even technology – is what business outcomes do they want to achieve?

You can think of some, let's say, basic things like we want to have fewer people so we want to reduce labour. That's a business outcome we want. We want to have more people come into our store, so we want to increase foot traffic. That's a business outcome we want. We want to create a loyalty program and we want to have personalisation in the way we communicate with our customers. We want more customer engagement. If you're any kind of business ... a retailer, a distributor, a wholesaler, manufacturer, utility, a mining company, whatever, you've first got to define what you're actually trying to do from a business perspective. Then, as a secondary step, you look at what are the available technologies and in addition to technology, what might have process changes, what might have people changes. So technology is not the be all and end all of creating a business outcome. But you look at what technology is available to me as an organisation, to generate that business outcome. AI is one of several things that might be available and may be the best approach.

Jarrod: That makes great sense. I think it comes back to the definition you gave around specific AI. Looking to solve that business problem, you may use a specific part of artificial intelligence to help gain that outcome you're looking for.

Matt: Or you might use something completely unrelated to AI to solve it. My father always called it the golden hammer approach. If you've got a golden hammer, then everything begins to look like a nail. Then you go in and start looking for nails and start hitting things, but not all business problems are nails. There are instances where, actually, to achieve a business outcome, you could potentially even change the process and change some of the people you have in the process, and you achieve the business outcome without any technology.

Jarrod: From your market engagements to date, what people considerations do you think organisations should consider when they're embarking on a digital transformation journey?

Matt: If you take an organisation, you've got, let's say, three sets of people. You've got all who work inside of your organisation, let's call them employees, contractors, staff.

You've got people who are your customers; you might have a business customer and it's the people inside that business, or your customer might be an individual consumer. Whatever, but those are the second group of people.

The third group of people are your owners, your stakeholders, and other interested parties. So if we forget the third group for a second, our shareholders, stakeholders, etc., and we just focus on our staff and our customers, I think you could cut them in half and say 'When you engage with customers, how can you make your customer experience, for lack of a better word, as delightful as possible? How can you make it a joy for customers to engage with your organisation, to buy from you? How can you educate your customers? How can you deliver more value to your customers?'

In this group, or this people box that is external to you, if you are actually serving, it's all about making their life easier, creating efficiencies for them, helping them succeed, giving them a great experience, educating them, and so on. It is a variety of different techniques, technologies, applications within a digital strategy that would be aimed at doing that.

Then, if you go and look at the people inside of your company – employees, staff, and so forth – you want to create an organisation with as much efficiency as possible. You make sure that ... like the principles of Lean: eliminate wasted time, eliminate wasted personnel, all that sort of stuff. You want to use technology to augment the capability of the staff that you have to allow them to do the job that they do, whether it's an operational job or a client-facing job, in the most efficient manner possible, in the best way possible from a process standardisation point of view. Again, I put it into these nice neat boxes when you talk about people.

Jarrod: I think really, in summary, it's all about improving the customer experience and improving the people experience inside your organisation.

Matt: Exactly. My philosophy is that as entrepreneurs ... the customer's the centre of their universe so everything is about the customer.

We've got to change as the customer changes, or as times change. We've got to look for more opportunities to improve what we do for our customers ... everything radiates out from the centre of that wheel, which is the customer, the hub.

Jarrod: Now ... moving onto the vendor landscape. The landscape has become increasingly crowded over recent years. How do you see the technology landscape evolving and what do you think the implications are from the customer's perspective?

Matt: There are a lot of big questions in there. First of all, the technology landscape will continue ... well, let me make the following analogy. Yes, it's become crowded and there's a lot of people doing a lot of things. However, put your mind back to when the Macintosh came out in, I think it was '84 or '85. That was just a breakthrough. That was just an amazing 'wow' moment. If you go today and look at that Macintosh, it's in an antique shop somewhere. Right? And you would laugh at its lack of computing power, at its graphic interface; it's just a relic.

If you look at it from that perspective, just put yourself 20 years into the future, that's to say nothing of 30 years, just 20 or even 15 ... and these great big systems we have today will be resembling that Macintosh. Things are going to be very different in the future and the things that we see as incredible today will become like the Macintosh. They'll be antiques, they'll be relics of the past. There will be new things ... new industries created with this; like driverless cars, whether it's holographic meetings and so on, that will spawn entire new areas of software development, add-on applications, etc.

I'm never really bothered by a statement like the technology landscape is crowded. Yes, it might be crowded today, but I'm always looking at what is this going to look like five to seven years into the future? What do I believe or what's my hypothesis in terms of what is going to be a trend? What problems remain unsolved, or versions take place, and how can you move towards being in that centre of the future hot spot as you manage to grow and evolve your business, if that makes sense?

Jarrod: It does. In terms of some of those new industries that you're thinking about, is there any that you would like to speak about that you think might be coming up on the horizon?

Matt: There's a lot of really interesting things that if you extrapolate them into the future have mind-boggling interpretation. 3D printing is a great example. To give you just a few data points. To repair a part on a satellite, it used to be an endeavour and still is today, but it's going to change quickly. You've got to launch a rocket into space, someone has to have that part, an astronaut. They've got to do a spacewalk to the satellite. They've got to go and take the part out that's being replaced. Then you put one in, etc., and come back to Earth. [It's a] hugely expensive and dangerous undertaking. Today, they actually have some satellites equipped with 3D printers that when a defective part fails, the 3D printer on the satellite prints it and then with a robotic arm it gets implemented. Right? From that, they've also now printed 3D edible food. Like Star Trek, you've got a computer making a pizza and it materialises.

Also, another current data point: in China, I think the first two-storey house that you can live in has been 3D printed. If you look just 30 years into the future, you could have an entire city 3D printed – skyscrapers, etc. That will change the goods and materials industry, it will change the supply chains industry, it will change the soft windows industry. You look at just this one technology like 3D printing. You can look at a whole lot of other things like internet of things, connecting machines to the internet, remote sensing, controlling the machines in response to senses and so forth.

When I look at these various areas and just extrapolate basically, not in any kind of exponential way, but just basically, the development in those industries and what it would mean, what impact it would have, it will turn technology landscapes upside down that are applicable to those industries, and suppliers supplying those industries as well. I think we're in for a lot of change in the years to come.

Jarrod: My final question. Matt, what is your greatest hope and your greatest fear in relation to AI?

Matt: My greatest hope is that general AI will never be fully realised in the sense that ... actually, what I believe, as we peel back the onion of the human brain, which is not fully understood today, the human brain is far from being deciphered, documented, classified, and it's like a diagram. It's a mysterious thing. Stephen Hawking said, 'The human brain is the most complex thing ever created in the universe'. So my great hope is that we will never fully understand it and, as a result, we will never fully be able to replicate it in a machine. Because I think if we do unravel it and we replicate it in a machine, it takes, in my view, all the beauty out of life.

It shows that total biology can be reduced into code and all we are is really a biological machine and we can create a digital version of it and the digital version is going to be a million times more effective; it will never get sick, never die, never make a mistake. All the kind of complications that go forward from that; how do people work? What is really even the purpose of a human being if you can just create a machine that will do any job better than anybody? Then what's our purpose? Forget companies, but just as a species. The purpose of a cow might be to put beef on the table and create milk and that's a great purpose, but what's a human's purpose? Is it just to be a consumer? Is it just to destroy the resources of the land? That's my greatest hope: that biological aspects of *homo sapiens* will remain mysterious, un-deciphered, and to some extent, un-replicable inside a computing environment.

And my greatest fear is that that won't be the case.

Jarrod: That's a great answer and it's one I concur with.

Matt: The more I feel as I go through life, and the more technology is around me, I always feel like I'm living less. I'm experiencing less. This is just my view. When I was a kid, my experience of life was the wind, the sun, the bicycle, the riding down the street with my friends. Then I look at my kids now and that's not their experience. It's video games and online and Skype, etc. Maybe I'm just getting old and sounding like one of those old people, talking about how the new generation is different.

My view of that is that they are experiencing less as humans than how I experienced it and so forth. I think the further we get ahead with the technology, maybe the experience becomes more and more limited until it becomes totally virtual and disappears all together. Those are the things I worry about when I have time at night before falling asleep.

Jarrod: I think you're totally correct in your thinking. Matt, is there anything else that you'd like to add from your perspective that you think is relevant to our reader?

Matt: In that space of workforce management and human capital management, and AI and automation, there's a really interesting thing in that people are connecting innovation and AI and making a direct link to the workforce. They're saying AI and all this innovation is bad because it's going to eliminate jobs and so forth. In some cases, they might be right in how we've created this technology that eliminates a job type all together, and it's not replaced by anything else because there are just too many people on the planet. But let's leave that argument to the side.

The problem with that model is, on the left-hand side you've got technology and AI pointing an arrow towards labour and saying this is having a negative impact on labour. But it's missing the most important part in that equation. That is capitalism. It's not AI that's driving these disruptive changes, it's not even innovation, it's none of those things - it's capitalism. People don't understand that the way the capital market works, we're not in a socialistic society, we're not in communistic society - we're in a free market capitalistic society, which means that if I'm a capitalist, and say I'm a hired CEO of a big company or I'm on the board of directors of a big company or even a small company, then my legal duty - this is even if you look at the ASIC guidelines - my legal duty is to value the shareholders. My job is to maximise the value of that company, protect it from risk, make good decisions and so on. Basically, we're operating in an environment where every company is being driven by people who are trying to maximise the value of those companies for the shareholders, and trying to make the shareholders richer. So then this should come as no surprise. That's how dividends are paid, that's how the business world works.

However, the problem is that these people who are trying to fulfil these managerial directorship duties and maximise shareholder value, they view labour as one of the most obvious targets for increasing shareholder value. The conversation will go like 'How can we reduce our sales force by 10 per cent?' Going back to one of your first questions: How do we use AI? It's not about using AI, it's about the business outcome: my business outcome is I want to cut my labour by 10 per cent. How do I do that? Here's an interesting technology, AI, and maybe it could help me eliminate people. You make this connection, AI eliminating jobs, but that's not what's driving the elimination of jobs. It's capitalism.

Because if I eliminate the jobs, I make more profit as a company, and that increases shareholder value. The company becomes more valuable. If I'm an executive, I'm likely to make my bonuses then.

Entrepreneurs are sitting to the side and saying 'Hey, these big companies want business outcomes like cutting labour, so I'm going to create an AI company that will help them eliminate the labour, help them achieve their business outcomes, and if I do a good job by building a good piece of software and delivering on my promise, I'll become a billionaire in the process.' That's a capitalism driver as well.

I think what is really driving all the change in the world today is not the goodwill of humanity or corporate social responsibility, it's none of those things ... What's driving all of this today is capitalism. It is how our society is set up. That is, in my view, one of the most interesting things to explore. And where will that lead us in the end?

Jarrod: Great views and great thinking. That's very thought-provoking, a jolting reality that people can take into consideration.

Matt: Innovation and AI are driving change in the workforce and you can explore that. The thing that's not occurred to us is what's driving the innovation and the AI? That is a really fascinating subject. Food for thought.

Jarrod: Thank you, Matt.

Matt: Pleasure.

* * * * * *

Top take outs

- → We are all part of an interconnected digital network and will transact on this basis into the future.
- → Remote work, team-based work and self-management will continue to become more prominent.
- → Some jobs will be replaced by workforce robots, but a human will still be required to complete them if the workforce robot is out of action.
- → Machine-based learning algorithms will continue to become more intelligent, taking omni-channel data as input.
- → Organisations will need to address the impacts on workers; some roles might disappear, while other roles might change considerably and require reskilling.
- → AI can be considered in two categories: *general* AI where the human is fully recreated by a machine and *specific* AI where specific problems are solved by machines.
- → It's not AI that is impacting on jobs, it's capitalism, according to Matt Michalewicz.

Where to next?

I hope this chapter provided you with an understanding of how technology and AI is playing out in the world of WFM. Matt's interview posed some interesting considerations around the capitalist society in which we live and how this may be accelerating the use of AI. In the final chapter of Part I we look to the future of WFM and how it might be delivered in the fourth industrial revolution and beyond.

---

1 World Economic Forum, 'We have the tools to reskill for the future. Where is the will to use them?', January 19, 2018, viewed February 18, 2018

Chapter 5

THE FUTURE: HOW WILL WFM DELIVER AND INTERACT?

We often find ourselves discussing the future with family, friends and colleagues, whether it's on social media, while watching television, or just chatting at the coffee machine or over a meal. It always leads to the same question: where will the future take us? There is no definitive answer. In WFM, however, there are a few trends emerging that enable me to make some educated predictions.

This chapter is a 'helicopter view' of where I see the industry heading, looking at technology, digital, adoption, delivery, standards and the greater good.

In some of the examples I provide, there are actual situations where suppliers are providing these functions now.

Technology

First, let's look at the largest contributor of advancement - technology - and the areas of most interest within this category.

Transition to cloud

The transition to cloud is expected to continue at a rapid rate. There will be a race for vendors to provide multi-tenant solutions that are scalable and elastic, so that customers and vendors can leverage efficiencies associated with this. Some WFM vendors have already achieved this outcome, while others are still moving to this space. It is important to note that just because a vendor has emerged with a new cloud platform, it doesn't mean they are the right vendor for you. Traditional considerations such as size of vendor, its ability to deliver, ability to support, how much it invests in research and development (R&D), the client base, should still be investigated before you choose a vendor.

On-premise

In my view, on-premise solutions will fade over time, but some customers will still prefer access to this option for an extended period. To stay competitive into the future, cloud technologies are undoubtedly required. There is a long runway for some of the larger WFM on-premise products that are entrenched into enterprise (large) customer space, as the rate of change in these organisations can be slow. As an alternative to an immediate cloud move for these enterprise customers, the vendors are offering hybrid cloud solutions for these products. For example, they are putting an on-premise product into a cloud-hosted environment and moving the responsibility to maintain the technology layer back with the vendor.

Point solutions, integration, solution extensions, marketplaces

A plethora of targeted cloud-based point applications already exist or are appearing on the landscape. Examples include specialist applications and services for onboarding, contingent worker management and learning. These are available as solution extensions via marketplaces. We have also seen organisations emerge that specialise in linking these systems together via cloud-to-cloud, cloud-to-on-premise, etc. The challenge here is to ensure that if multiple applications are adopted into an organisation, the end-user is not forgotten, the solution is usable, and the business objectives can be met. The experience should be 'frictionless'.

Applications that are predominately self-learnt are becoming normal. It's also becoming more common for users to work across multiple connected apps such as Facebook, Instagram, your favourite email solution and so on. Business applications are no different.

WFM solutions as a service

Organisations that offer WFM solutions as a service (e.g. roster to pay) are now available. These organisations offer the solutions as a holistic service where the provider takes responsibility for the overall solution including the infrastructure.

Algorithms as a service

Specialist organisations are writing algorithms and using mathematics to solve specific business problems. Some examples include:

- → Customers come to a store and queue up behind three cash registers. Specific problem: what is the optimal way for customers to queue, to minimise their wait time but also to minimise the number of cash registers required to be open?
- → What is the optimal balance of full-time versus casual staff to minimise the cost of labour while maximising time spent with a customer?
- → In a warehouse, what is the optimal way to minimise the number of staff required to fetch goods and to minimise the time associated with fetching the goods?
- → What is the optimal way to work out the quickest route for a healthcare worker to service home-based clients and also minimise the worker's kilometres travelled?

These specific business problems need specialist optimisations that have many variables based on forecast or actual data which drive the answer.

As organisations continue to drive efficiency and productivity improvements, there will be an increasing demand for these types of services.

Platforms

Platforms are emerging that are open in nature and where you can access developers to build extensions that solve your specific business problems. Platforms have an open application programming interface (API), which enable extension development. The benefits of these platforms are that they provide fast innovation, are on current technology, offer an ability to scale and solve specific problems by creating a solution using available information.

Let's take a look at an example: a retail store may be looking to better align over- and under-staffing in real time. Data can be gained from multiple sources, which influence forecasts such as event information and foot traffic. Algorithms can be used to re-forecast in real time and update revised forecasts automatically to the platforms via open API.

Technology commoditisation and consolidation

The market is filled with many new, capable and disruptive product suppliers. Workforce technologies currently being developed are highly usable and designed to meet the needs of not just big companies, but also small-to-medium businesses (SMBs). This recognises that the digital workforce is becoming pervasive in every part of the economy.

These product suppliers are working furiously to develop and refine their tools and templates, to ensure implementation and configuration processes require the lightest of touches. This light-touch approach allows the product suppliers to focus more closely on differentiating their product by functional area, customer or industry perspective. What this equates to is specialisation and personalisation, both key requirements in today's business environment.

It's important to note that many of these vendors have large investors and/or venture capital companies backing them. In an increasingly SaaS-dominated market, this has resulted in a race to acquire more customers so that investors can realise their return on investment. While selling seats has always been a high priority with product suppliers, the stakes are now higher, as more suppliers look to secure a position in an already congested market.

Despite the explosion of products in recent years, few suppliers can truly meet the needs of enterprise clients by providing items such as account management, global implementation and support, multilingual functions and industry expertise.

I see product suppliers consolidating rapidly in the near future. Those product suppliers that will survive the 'cull' will undoubtedly be the ones who can be the quickest to adapt to their current and prospective customer and industry needs.

Digital

Next, let's look at the future in the digital space.

Big data/data intelligence

A massive opportunity exists to tie detailed operational WFM data to data from other systems such as HR, sales and finance systems. The move to cloud systems is helping this to occur as the data from multiple customers is accessible in a collective manner.

In the future, we will be able to generate industry-focused statistics and benchmarks. Here's an example.

Suppose the absenteeism average in your business is ten days per annum, and the industry best practice is eight days per annum. Your business is running two days per annum behind the industry best practice.

Data intelligence across multiple data sources will enable your management team to identify the answer to why your absenteeism is running higher than the industry average. Management dashboards will be available where metrics and KPIs can be set based on the results of the data intelligence activity, along with the inputs from industry best practice. These dashboards will continually update based on real-time data collected across multiple data sources. Businesses will be able to see how they are tracking against their metrics and KPIs, and be able to continually identify absenteeism trends and how they are tracking against others in their industry.

There are privacy considerations here; particularly when bringing in multiple customers, operational WFM data is detailed at an employee level. No doubt the implications will set a new challenge for WFM compliance specialists to solve.

Predictive and prescriptive analysis

There are significant opportunities to tie operational WFM data into supplementary business systems such as HR, sales and finance. WFM data is at the lowest level of granular detail associated with an organisation's people. For example, we can already ascertain when a person arrives and leaves work; what time they actually start and stop working, what jobs they are doing, or how long they took to do the job.

Using WFM data, we will be able to predict:

→ how long a job will take;

→ what type of people profile will get the job completed in the quickest time to the highest standard;

→ the best team composition to work a particular shift; and

→ employee performance to proactively address in periodic reviews with their manager.

A simple example of this is the ability to automatically generate entries in the HR system to remind managers to speak to employees who are habitually late. If the issue persists, the system could automatically generate a performance review record to discuss their lateness, which could then also be used in a salary review discussion.

The WFM system could analyse sick leave data and proactively alert a manager that a particular employee is likely to take a sick day. The system could notify the manager suggesting the employee take annual leave instead. This will reduce their sick leave balance, minimise the chance of the employee getting sick and improve employee satisfaction because their employer is caring for their wellbeing in a proactive way.

Certain industries are dependent on the weather. For example, a cinema will have more patrons on a rainy day because there are fewer options for outdoor activities. The WFM forecasting system could check a weather feed to dynamically ensure an increased number of staff are rostered to service the demand.

Predictive and prescriptive analysis will utilise AI, machine learning and deep learning to deliver optimal outcomes.

Self-learning systems

Future WFM systems will learn what functions a business requires, based on characteristics such as time of log in. For example, a manager may log in each day between 7:00 and 9:00 am and complete the same task. The WFM system could predict the manager will repeatedly do this and set it as a default. It could also learn that the first task completed is to replace those workers who are sick on a given day. The system could automatically determine which alternative workers are available that day and propose they cover a shift.

The software could ask people simple questions each day. For example, do you have spare time today? Are you in a good mood? Based on their response, the system could alert fellow workers to be cognisant of this. The software can also assign specific tasks based on this information.

We can see how self-learning systems will utilise AI, machine learning and deep learning to deliver optimal outcomes.

Wearables

While not a new concept, wearables will allow more intelligence, such as responding in real time to data triggers. For example, a wearable could vibrate when you need to take a break, or monitor the length of time you have been working on a job. A worker will be able to move around their work area and the technology will be able to determine information related to location, cost tracking, etc. These wearables could become integrated with work attire such as be embedded into work boots or hi-visibility vests.

Communicative interfaces and robots (BOTS)

I vividly remember when I completed my first computer science degree 25 years ago, hearing about the possibility of AI impacting certain areas of life. At the time it seemed like the stuff of science fiction movies; today, we see practical examples of this everywhere.

We already have examples of communicative interfaces available in the public realm with technologies such as Siri, Google Voice, etc., but we will see a large-scale introduction into WFM systems. A worker will be able to send an email or provide a voice prompt, for example: 'Find a replacement worker for my shift this Friday'. The command will trigger a set of behind-the-scenes functions to find a suitable replacement (potentially a contingent worker), complete the necessary approvals and automatically replace the worker for their shift.

Another example may be triggered via a manager: 'See if Joanna can start her shift half an hour earlier tomorrow. If Joanna can't do this, use the call list to cover the half hour based on minimum cost'.

Multi channel and omni channel

Up until around the time of the third industrial revolution, most interactions across industries were physical. You went to the store to buy an item or to the bank to deposit money. From the 1970s onwards, additional channels started to develop. The call centre and email were prominent interaction channels for a long time.

We are now seeing the rapid development of multi channels by which we interact, for example via social media (Facebook, Twitter, What's App, WeChat, Instagram), online chat, voice and specialist applications to name a few. Coupled with this we live in a 24/7 'always-on' economy where our employees and customers are looking for instant and seamless answers to their questions across any interaction channel they choose - iPad, tablet, PC and so on. This interaction is delivered by integrated omni channels that provide a unified and 'frictionless' experience.

Organisations will look to enhance this experience by partnering with other organisations that can deliver technologies to support this frictionless, omni-channel experience or business functions that can be plugged in to provide value-add functions to solve specific business problems or to enhance loyalty to their brand.

As an example, let's take a contact centre in an insurance company. Your knowledge of the customer will come from not only the policies they have with you but also from the data collected via IoT technologies - such as the amount of time they are in their home. You will be able to source data from social channel interaction and other data a customer chooses to share with you. Data science will examine this information to better align policy inclusions and risk ratings prior to quoting a policy renewal. You learn from a customer's social interactions that they use Spotify to listen to their music. As part of the renewal you are able to offer a subscription to Spotify. Your business model changes to schedule your staff to proactively arrange a time to call the customer to review the policy renewal with them and determine the appropriateness of the Spotify subscription reward. You now require a role in the contact centre related to outbound customer retention. As part of the retention process, the customer wants to review the policy with you at a time that is convenient to them. The call agent works from home to better align their availability with the specific customer-driven call times.

The move to omni channel and frictionless employee and customer experiences will grow and become part of your core business operating model. Forward-thinking ways of work will be required to remain current and competitive.

Adoption

Technological and digital advancements are all well and good, but how will they be adopted in the future?

WFM and HR will combine to create a 'complete people picture'

WFM and HR will be required to merge to complete the big picture and to improve employee engagement and experience. They will have to work together to understand each other's purpose in bridging the gaps between senior management, HR, operations and the broader business value chain. Agile methods will be required to prioritise, educate and deliver these combined initiatives, enabling focused and accelerated delivery of business benefits. Digital transformation will be a key requirement.

This combined people function will play a role of utmost importance, as people in organisations are at the epicentre of purpose and culture. To deliver a complete people picture, WFM and HR functions must be considered together to derive people's maximum value. Examples of the combined benefits are represented in Figure 5.1.

Figure 5.1: WFM impacts in your business – your complete people picture

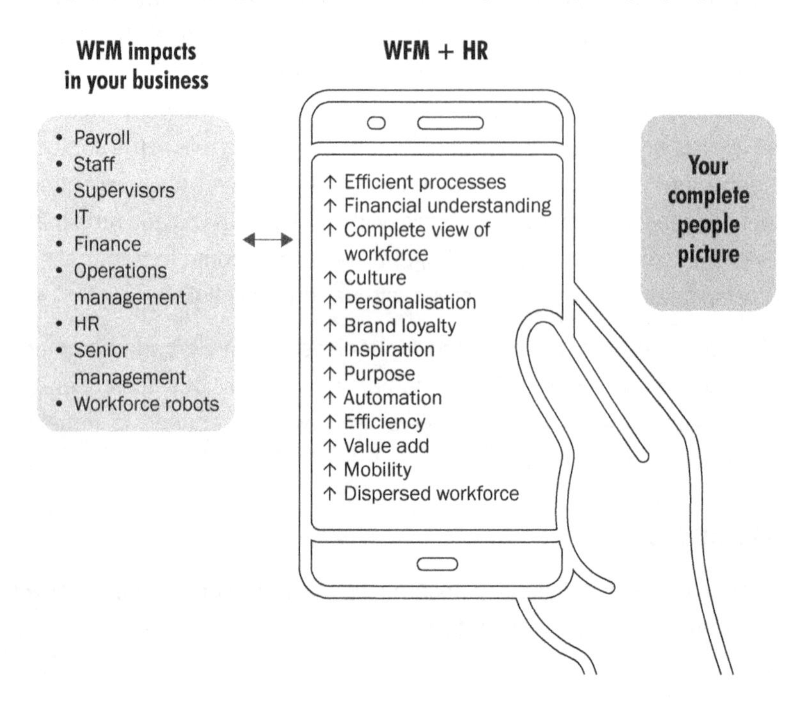

Strategic business operations

We will see the introduction of specialist industry operational teams augmented into business executive teams to drive operational improvement. These specialists will be led by seasoned professionals with many years' experience. The teams will leverage extensive business operational knowledge aligned to optimal use of WFM technology. Examples in a health industry environment may include directors of nursing and nursing unit managers, and these specialists will know how to strategically enhance business value.

The concept of the portfolio worker, which Georgegina Poulos introduced in her interview, will also be relevant here. Employers will be able to tap into the right skills on flexible terms to solve specific business needs.

Delivery

Once adopted, how will the advances in WFM be delivered?

Niching across suppliers and talent

Niching involves providing a specific mix of tools and skills to derive greater value.

Software vendors will further niche their technology functions to support industry specifics while adopting the latest technology advancements, such as AI, to support this.

Talent to deliver and support these initiatives will become concentrated in global pools, allowing service staff and clients to leverage knowledge in a far more collaborative and simplistic way than we have experienced previously. Industry knowledge will also have a major influence in these talent pools.

Service-providers will supplement specific WFM talent into client-side roles to provide or support business operational functions. For example, a WFM administrator will ensure operational managers complete their sign-off prior to payroll extract.

The rise of team-based collaboration

In recent years, we have seen a focus on talent acquisition and talent management. As we move forward there will be a greater focus on remote work, team-based work and WFM. This will enable business outcomes to move from focusing on systems of record to systems of productivity. It will also provide a great opportunity for managers in organisations to further develop their management and leadership skills, while their team members will receive improved communications and focus on the outcomes they are required to achieve.

Productisation and consumption

Increasingly, we will be able to buy software and services as products. The way in which we buy will become very consumable, analogous to the way we buy from an e-store today; we look at the range of products, quickly assess which one is suitable for our needs and click to buy it.

I also see that just as you buy software on a dollars-PEPM basis, you will also be able to buy services in a similar way. They will all be available from a supplier price list. For example, you'll be pay $X per month for software and $Y per month for services.

Outcome-based compensation

As knowledge of industry gains and niching continues to grow, confidence to deliver outcomes will increase, and vendors and service-providers will start offering organisations outcome-based fee-payment options. For example, based on industry data, using a defined process we can increase sales by X per cent and decrease costs by Y per cent. They will guarantee this, and payment will be fixed based on a defined outcome.

Another example could be related to client augmentation, with an executive level strategically focused business operations team, providing operational efficiencies that can be defined to increase sales revenue by X per cent. In return, the service-provider will be compensated with a fixed payment and a share of the increased revenue or profit.

Standards

The WFM industry has been slow to develop any formal degrees or qualifications. We are, however, seeing more structure, standards and expectations appearing globally, which will see the introduction of this specialist skill at academic levels.

We will see more predictive and proactive management of talent. WFM will start to become more formally recognised, with stronger customer advocacy and a greater understanding of the critical role WFM plays in delivering consistent business performance.

We will see the introduction of formal standards. These may come via organisations such as the International Organization for Standardization (ISO) and will become increasingly important as WFM permeates more areas of the business world.

The greater good

Finally, how will advances in WFM affect the greater good?

Increasingly, organisations will become focused on what the real purpose of their business is. Is it solely to make money? Are there other reasons? Organisations that have a greater sense of purpose to support the environment, the community, social causes or sustainability will be in a stronger position to attract and retain the talent of tomorrow.

Wrapping it all together

In the short term, WFM delivery options will continue to grow, leveraging different platforms and increasing the levels of service provided via PEPM pricing.

In the medium term, we will see more predictive and proactive management of people. Businesses will have access to highly skilled industry specialists aligned to technology, who can drive operational improvements at the senior management level. WFM will start to become more formally recognised, with stronger customer advocacy and a greater understanding of the critical role WFM can play in delivering consistent business performance.

In the longer term, we will see HR and WFM combine to provide a complete people picture. There will be practical use of technology as workers become increasingly self-responsible. Employers will move from a reactive to a proactive approach to WFM. Big data and data intelligence will provide tangible metrics, allowing KPI alignment to industry. There will also be a move from PEPM pricing to a results-focus on defined business outcomes.

Whatever changes or technological advances are ahead for WFM, I strongly believe the future is bright. As WFM becomes more mainstream and its sphere of influence grows, businesses, WFM practitioners and technologists will continue to push the boundaries of what is possible, to deliver unparalleled value back to the businesses, people, customers and society.

Top take outs

- → Service industries will remain relevant to assist organisations in navigating the future successfully.
- → Teams will be highly niched and spread across the globe, focusing on business outcomes.
- → Business outcomes will be delivered by collaboration and partnering across many providers.
- → Machine learning and data intelligence will continue to advance and provide greater opportunity for prediction, prescription and automation.
- → An omni-channel approach will be utilised to provide frictionless outcomes from employees for customers.
- → Formal industry certifications will become more prominent and focus on the solution to business problems.
- → Compensation for driving value into organisations will be based on outcomes.
- → Organisational purpose, outside of pure monetary value focused on society, will become the norm.

Where to next?

That concludes Part I of the book. You have now learned about the evolution of WFM, what WFM is, the benefits your organisation can realise from WFM and the flow-on effect to create better people experiences and better customer experiences. We finished with a look at where AI is taking WFM and future trends from an interaction and delivery perspective.

Part II starts with the ground rules and fundamentals you need to ensure the success of WFM initiatives in your organisation. Then we move onto looking in detail at the 5-Step Methodology to Smarter WFM.

PART II

THE 5-STEP METHODOLOGY TO SMARTER WORKFORCE MANAGEMENT

'If the rate of change on the outside exceeds the rate of change on the inside, the end is near.'

Jack Welch

Chapter 6

THE FUNDAMENTALS TO ENSURE SUCCESS IN ANY WFM INITIATIVE

Having had the privilege of working with many organisations on their WFM initiatives, I've realised the WFM investment process is highly predictable; the same problems occur, over and over. So, I've identified a number of key areas of the WFM investment process that, if enacted correctly, will ensure that the investments you make in WFM return the defined business benefits you are seeking. Following this methodology will help develop your workforce, tighten its culture, improve brand loyalty and strengthen alignment to long-term organisational goals.

From my experience I have seen many organisations, in particular product suppliers, focus on implementation. While implementation is an important step in the process, it is only one step that your organisation needs to undertake. Hence, the methodology I have created – Smart WFM 5-Step Methodology – looks at the *entire* WFM lifecycle. You can adopt the methodology at any of the five steps, moving back and forth between each step relevant to your business initiative, as depicted by the bi-directional symbol in Figure 6.1.

Top of mind in this methodology – however you enact it – is people.

Smart WFM 5-Step Methodology

By way of introduction to the Smart WFM 5-Step Methodology, let's take a look at Figure 6.1.

Figure 6.1: Smart WFM 5-Step Methodology

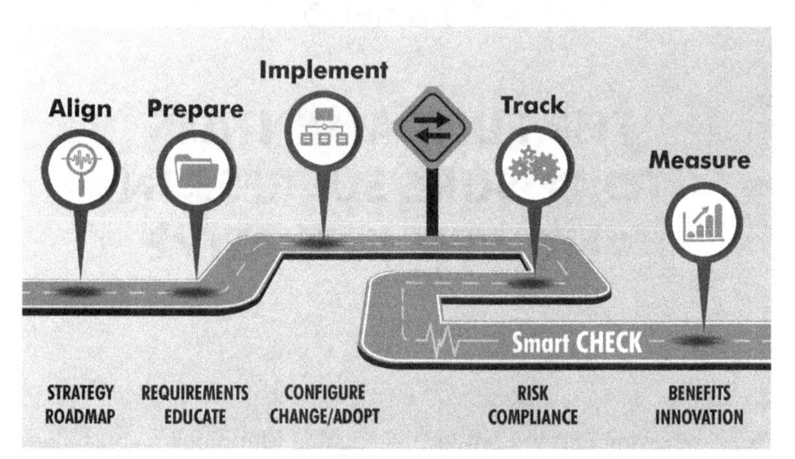

Each of the following five key steps of this methodology is covered in detail in a separate chapter, but let's take an overview of them now.

- **Step 1 – Align:** Make sure you understand your business strategy; align it to your workforce goals and create a baseline to achieve your WFM outcome.
- **Step 2 – Prepare:** Ensure your business is educated and ready for the transformation journey ahead. This includes confirming alignment between senior management, operations, finance, HR, payroll and IT, and understanding what business impacts will occur.
- **Step 3 – Implement:** Enable your people, processes and technology so your people know what to do and when to do it.
- **Step 4 – Track:** Confirm governance is in place to mitigate risk, ensure compliance and keep your WFM initiative moving along, balancing the inputs from your various business stakeholders.

→ **Step 5 – Measure:** Continually review the benefits to ensure you receive optimal value from your WFM initiatives.

Smart CHECK

At any time during the business lifecycle, doing a Smart CHECK can determine where you are at in your process and what you need to do next. The Smart CHECK can be completed by step, by specific tasks in a step, or any combination of step and task.

A Smart CHECK will generally look at people, process, technology and/or financials and deliver a gap analysis. I have included a framework for you to follow when you conduct Smart CHECKs in Chapter 12.

Tips for smarter implementation

A wise person once said, 'don't make your own mistakes, learn from mine'. So before we drill down into the five steps, here are some tips to make sure you avoid mistakes I have seen made in the past that make WFM implementation more difficult.

The importance of storytelling

In the past I have seen many WFM initiatives become consumed by technical and IT speak, which has prevented business stakeholders and senior management getting to the point quickly. When you present the challenges and benefits of your WFM initiative, be conversational in your delivery, speak in plain-English and avoid (or explain) jargon, tell a story. This will ensure complete engagement with your audience.

Keep the WFM analysis framework front and centre

You will remember that I developed an analysis framework to help you present the benefits WFM solves in a coherent way. This framework introduces the four main business problems WFM solves, and it's worth taking a moment to revisit these. This framework can be used for almost any challenge that presents itself during your WFM journey.

Figure 6.2: The four main problems WFM can solve

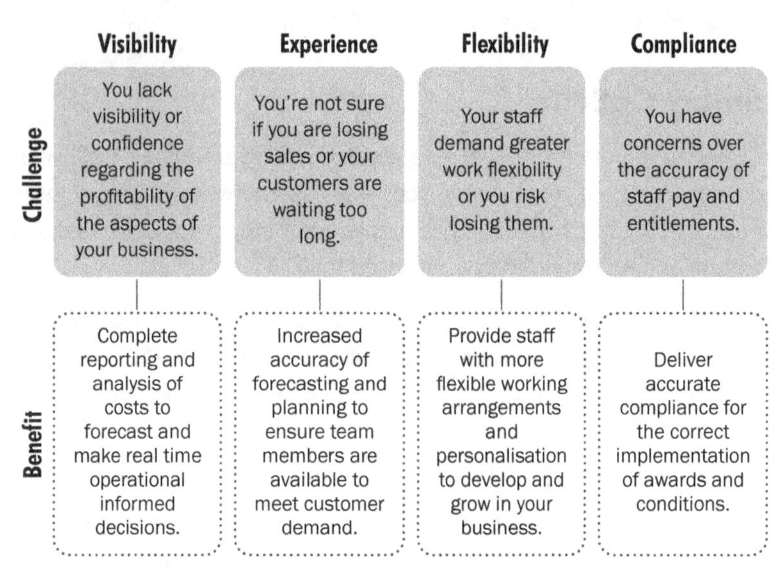

Remember the Smart WFM 5-Step Methodology is not a silver bullet

It's important to note that this methodology is not the answer for every business challenge you have; it isn't a silver bullet. Rather, it is a framework that can be applied to help guide you through your workforce-related business challenges. The methodology is designed to be used collaboratively with traditional methodologies and/or vendor methodologies.

Understand your organisational goals

Your organisation's senior management will be looking to achieve business success in several ways: by increasing the addressable market, increasing revenue, retaining the best staff, decreasing costs and/or reducing risk. WFM can help deliver on these business objectives. It will help if you prioritise these objectives and align your organisational goals to your workforce goals. Each time you make a decision, reflect on your organisational goals.

Make sure your workforce is engaged

Senior management, HR, operations, payroll, IT and finance must all be aligned to the business objectives. In addition, programs of work across these business areas need to align resources and outcomes to avoid contention and misunderstanding. Special attention should be placed on HR and WFM functions to make sure they are working together, as both are tightly coupled to business operations.

Define benefits and measures

Benefits must be defined in an easy-to-understand way. For example, if the business goals are to decrease labour costs and increase productivity, ask:

→ How will these goals be achieved? (Example: By providing tools to better align staffing to customer demand and making more time available for the manager to work on value-adding tasks.)

→ How will these goals be measured? (Example: By measuring labour cost pre- and post-implementation, aiming for a percentage reduction and the manager spending an additional number of minutes per day with customers.)

These benefits and measures must be tightly aligned to goals and, ultimately, business requirements. Step 5 of the methodology introduces a benefits and measures framework.

Understand the change, culture and mindset shift required to transform your organisation

In my experience, this is the biggest hurdle to overcome to achieve desired outcomes. Senior management are the key to achieving this and the greater the change, the more their involvement is required. As your workforce moves to a digital way of working, the underlying organisational core may need to change. Due to automation of processes, roles may change within the operations area of the organisation, such as: completing daily scheduling, taking responsibility for cost management, increasing staff interaction and goal-setting in line with organisational needs.

Don't bite off more than you can chew!

Focus on outcomes that align with your overall WFM maturity, the size of your organisation and specific localisations associated with your organisation, remembering that WFM is a journey, not a destination.

It starts with automation of time and the shift-based scheduling process. As your WFM matures, you can further automate the scheduling process, e.g. skills, least cost, using point-of-sale (POS) data to forecast. Some organisations are ready for this on day one; others are not. Be pragmatic and realistic.

Consider the end-user – frictionless interaction

If an electrician turned up at your home to install a light switch and they put it on the ceiling, what would you do? After your initial shock and disillusion, you would have them move it to the wall and likely never use that electrician's services again. Technology should be no different. Place significant focus on the end-user to improve the overall employee experience. Design-thinking approaches and user journeys are a great way to achieve practical outcomes with technology. Consider what you are designing from an end-user perspective, and create cases that define your requirements around this to ensure the solution is usable. Make the interaction frictionless.

Align the technology

Obtain IT alignment from the outset to ensure that your technology layer is aligned with IT standards (such as: service management, security, availability, support, integration and architecture).

Continual advances in integration are evident, with many products having an open API and out-of-the-box connections with complementary technologies, which play a major role in adoption and experience.

Get runs on the board quickly – agile and waterfall

More recently, implementation philosophy has moved from waterfall (fixed scope, estimated time, estimated resources), to agile (variable scope, fixed time, fixed resources). This is depicted in Figure 6.3.[1]

Figure 6.3: Waterfall and agile approaches

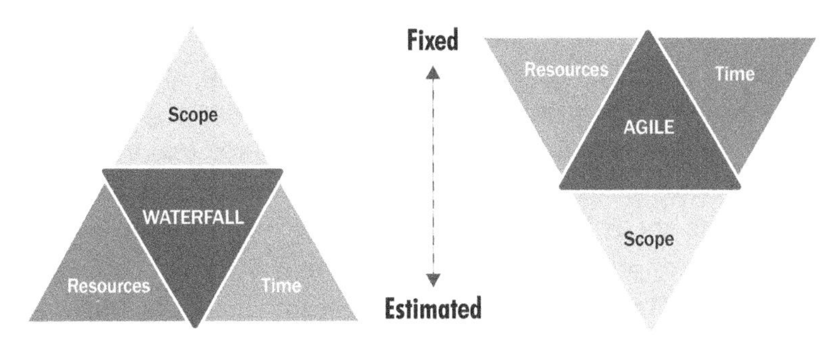

The key difference between waterfall and agile is what is fixed versus what is estimated. Agile projects have a fixed schedule and resources where the scope varies. Having fixed resources keeps the team focus tight as knowledge develops within the delivery team.

The agile concept lends itself well to delivering outcomes where you can continually prioritise and check against your business objectives, benefits, adoptions and employee experiences. I believe the agile implementation philosophy will become the norm when delivering WFM projects.

Have the right team in place to ensure success

To achieve greater value in WFM, the right team is crucial. This stands true for your team, your product suppliers and your service providers. Make sure you choose your team in line with what you want to achieve and how you want to achieve it. Ask whether your team members are:

- → quick to adapt;
- → good team players; and
- → comfortable dealing with ambiguity.

If you have a strong governance layer in place to manage the process from the outset, you will be in a stronger position to achieve the desired business outcomes. All parties need to have a seat at the table to collaborate and achieve the agreed outcomes.

Measure and innovate

With on-premise WFM implementation at the end of the process, the team generally disbands and that's the end of it. If this happens, then you have missed a vital step. Benefits need to be measured and value of investment needs to be understood. In the cloud world, we are in a state of continual new releases and enhanced functionalities. Make sure these functions remain aligned and measured against your business objectives. This can facilitate a mindset of continuous improvement resulting in productivity increases in your business.

No doubt there will be further advancements in technology and business thinking. As this occurs, we must adapt our approaches to keep pace, as the path to achieving greater value from WFM using digital technology continues to gain clarity.

Top take outs

- → The process to value is predictable, which is why developing a methodology made sense.
- → Don't overcomplicate what you are doing. Always ask yourself, what are the business problems you are looking to solve, and what is the benefit of solving them?
- → Follow the 'Tips for smarter implementation' given in this chapter when making decisions and adjust them to suit your business as required. Remember to consider all the tips for smarter implementation collectively.
- → Make sure you have the right team in place to deliver the outcomes.

Where to next?

Now I've introduced the Smart WFM Methodology, it's time to look at each stage of the 5-step methodology in detail, unpacking the key considerations to achieve value from WFM in your business. Are you ready?

1 T. Aljaber, T. 'The iron triangle of planning: The ultimate balancing act and how to achieve agile project management nirvana', Atlassian, viewed [2 December 2017], <https://www.atlassian.com/agile/agile-iron-triangle>.

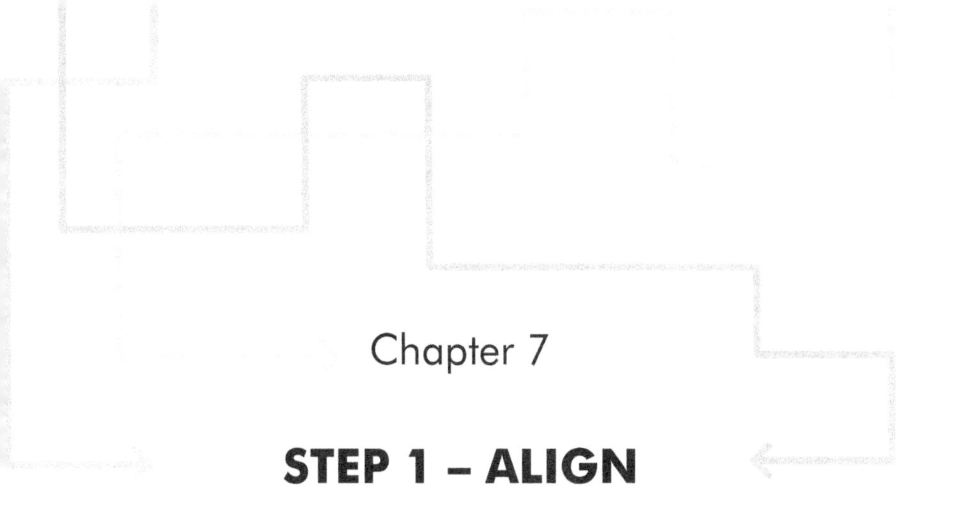

Chapter 7

STEP 1 – ALIGN

In the numerous WFM implementation initiatives I have been involved in or learned about, sometimes the CEO has had a clear vision of the business outcomes they wish to achieve, such as increasing sales. At other times, the execution of these initiatives has been completed simply to solve an operational issue such as to automate the collection of time.

Bear in mind that if your senior management signs up to an increase in sales and your delivery team delivers a new time collection method to support payroll, your initiative will not be successful. It's important everyone is clear about what you are looking to achieve and how you are going to achieve it – from the top down.

How to get value from WFM

Step 1 of the Smart WFM Methodology starts by taking a close look at the benefits you can expect to achieve from WFM. Once you understand these benefits, you can tie them to your business outcomes and create a strategy and roadmap to achieve this. We will also examine some of the considerations that are specific to WFM, enabling you to achieve value from an investment in WFM. This will, in turn, mitigate risk in any future steps you take to achieve that value.

Figure 7.1: Smart WFM Methodology: Step 1 – Align

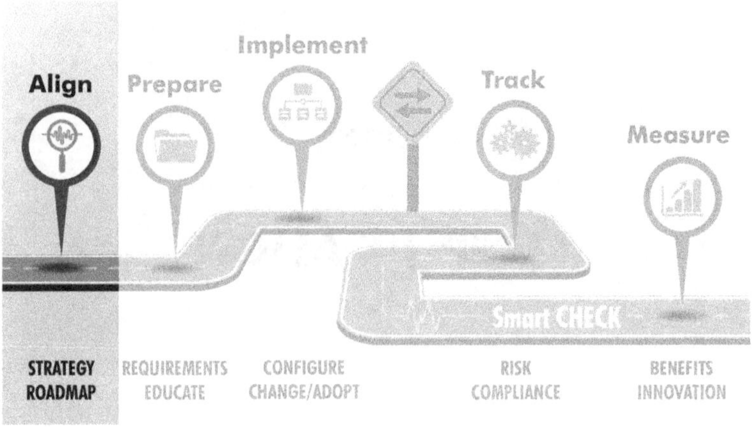

Benefits of WFM

Let's recap on Part 1. What is Workforce Management? We found that WFM influences many parts of your workforce from a people, process, technology and financial perspective.

Now we look at the key areas where your organisation can benefit from WFM. Each of these areas may be considered standalone, or there may be interdependencies with other areas. I have used a number of practical examples across a variety of industries to help explain this. While this list of the benefits you can achieve in your business is not exhaustive, you should be able to relate similar benefits to your workforce-related initiatives. You will also need to align the terminology to your specific industry.

Real-time visibility of people and costs

WFM allows for real-time collection and dissemination of information.

From a collection perspective, WFM can determine who has arrived for work, what jobs they are working on and where the work should be costed.

From a dissemination perspective, WFM allows staff to know where they are scheduled to work, what time they start and how long they will be required to work for.

> **Example:** Daniel is a nurse. He uses his mobile device to determine what time he starts work and what ward he will be working on today. Michelle, who is a nursing unit manager, knows that Daniel is at work and on the appropriate ward. Michelle is also able to track the actual cost of her team against the hospital department's budget.

Compliance

WFM is able to automate and build rules to manage compliance.

> **Example:** Allied Manufacturing Company creates an enterprise bargaining agreement (EBA), which sets out the payment rules and break rules for each worker type. The application of the rules can be automated, so the company knows that the payment and fatigue rules are being applied as expected.

Accuracy

WFM creates an ecosystem that enables accurate recording and approval of time.

> **Example:** James works for a council and his job is to repair damaged roads. James works as part of a team and his team knows which roads they need to fix throughout the day in real time. James's manager can allocate his team to the associated tasks required to fix the road. The costing information is then used by the council's operations team to understand the actual people cost associated with fixing the roads.

Full visibility of labour force

For you to manage and forecast your people, you require complete visibility of the workforce. Consider each of your different worker types: full time, part time, casual, contract, seasonal and salaried.

Many organisations pay close attention to the mix of these worker types. There are factors that influence this form of analysis, including the payment conditions associated with the EBA and how they relate to the worker type.

> **Example:** Jonathan changes the mix of full-time versus casual staff rostered to work on a specific shift. This change in staffing mix impacts the costs of the shift. Based on this, Jonathan is able to understand the impacts of these costs in real time and make informed staffing decisions.

Forecasting

Once you have timely and accurate people costs, you can use this information to forecast.

> **Example:** Melissa is the care manager in an organisation that provides health-based home care services. She can use the actual data to forecast her people costs for the next month.

> **Example:** Damien is a store operations manager for a large retailer. Damien uses historical sales data to forecast the number of team members required to work in a certain department. This allows accurate people-cost forecasting.

Many other factors can also be used to forecast. In retail, this may include foot traffic throughout the day. In health services, it may include patient admissions data. In a contact centre, this may include expected call volumes due to an event.

Efficiency

A review of your people, process and technology to support WFM enables you to drive efficiency improvements.

> **Example:** Before the implementation of a WFM system in a retail store, associate Jenny received her task list of work to complete at the start of a shift from her manager, Jeanine. If there were changes to the task list during the day, Jeanine would need to find Jenny and verbally communicate the changes to her. With the introduction of a WFM system, Jeanine can manage all her tasks for all her associates electronically. When the store gets busy, Jeanine can quickly reprioritise and communicate the tasks electronically, making the process more efficient.

Productivity

Improvements to your people, process and technology supporting WFM enables you to drive productivity improvements.

> **Example:** Before the implementation of a WFM system, the operators on a manufacturing line had many manual intervention steps to record start and stop times. With the introduction of WFM this process was automated. As a result, the production quantities increased.

> **Example:** The Apple County contact centre uses a workforce robot to take the first line of calls in its contact centre. The workforce robot is programmed to continually learn from the questions it receives and the answers it gives. If 90 per cent of callers ask, 'What time does [X] event start?' the robot can automatically provide this information whenever a call is answered. If the volume of calls increases, the workforce robot can clone itself and take multiple calls at once. As a result, calls are answered and handled quicker.

Culture and purpose

WFM allows your people to free up their time so they can concentrate on value-adding tasks to grow and drive company culture.

> **Example:** Store operations manager Gary knows he has the right people with the right skills scheduled to work. This allows him to spend time with his people to educate them on the company's organisational purpose, which was defined as part of the company's strategy as a key contributor in building company culture.
>
> Gary is also able to spend time with his people to communicate the tasks they can undertake to improve customer experience.

Employee experience and engagement

Many factors influence employee experience and engagement, and WFM is at the heart of this. WFM helps to make operational processes in your business more accurate, productive and efficient.

> **Example:** Molly, a retail associate in a department store, is able to use her mobile to set her availability for work and to acknowledge the shifts

she will be working. Molly is able to specify her availability for work as 9 to 5 each weekday and confirm shifts that fall in this timeslot.

A report in *Harvard Business Review*[1] found that inspired employees are three times more productive than dissatisfied employees, yet only one in eight employees feels 'inspired'. Knowing that your employees have less operational stress provides a platform to increase their engagement and make their work experience better.

Customer experience and engagement

Likewise, many factors influence customer experience and engagement, and WFM provides a key ingredient: people personalisation of the experience. Customer experience and engagement is driven by a combination of many of the WFM benefits; in particular, culture, purpose, employee experience, engagement and inspiration.

> **Example:** Retail associate Sarah is able to understand her customers' needs, provide customers with options, communicate her company's purpose and provide an overall personalised experience that is meaningful to her customers.

Customer experience and engagement helps differentiate your brand from that of your competitors.

Increased sales/service output

If you are forecasting accurately and providing superior customer experience and engagement, you are in a strong position to maximise your sales or service output.

> **Example:** Pauline, an associate in a retail store, takes the time to fully understand her customers' needs. She spends time with each customer looking for opportunities to upsell, which results in increased sales revenue.

> **Example:** James, a triage nurse in an emergency ward, takes the time to fully understand his patients' needs and ensures he accurately triages injury at first point of contact. This results in allocation of patients to the most suitable doctor in the first instance, resulting in greater productivity and servicing a higher number of patients.

Quality

Quality manifests from a number of WFM and broader benefits.

> **Example:** Nursing unit manager Michelle knows that her wards are adequately staffed from a people and cost perspective. Michelle is therefore able to focus on quality of care of her patients.

Ensuring there are measures in place is important to track effectiveness. Your organisation can use this as a key point of difference to create a quality focused mindset in your business and for your operational managers to take greater responsibility.

Leadership

WFM enables your organisation to develop leadership capability among your managers. In the interview with Georgegina Poulos in Chapter 3, she spoke about this in terms of T2 developing entrepreneurial traits in its managers, enabling them to take greater responsibility for their business areas. Once you embed this entrepreneurial thinking, your leaders can take greater responsibility for commercial and people-related decisions. This will positively impact financial performance and lead to direct benefits for employee and customer experience. Having visibility of rosters in advance, and the cost of these rosters, allows proactive decisions on staff to be made prior to the event.

> **Example:** Martin is a retail manager. He is able to make informed decisions about approval of overtime today. At the time of making his decision, Martin will be thinking, *how will this impact our profit and loss? How will this impact my customers' experience? What skills' development can I look to provide for my associates?*

Organisational goals

The first thing to consider when embarking on any WFM initiative is to understand your company's organisational goals. If you find your initiative is not aligned to the organisational goals, you need to ask yourself why this is the case. If you don't, a senior manager within your organisation is likely to ask you this question at some point.

Alignment of organisational goals to WFM goals

Once you understand your organisational goals, there will generally be company initiatives underway that you are looking to align with from a workforce perspective.

> **Example:** Currently in Australia there is a government-related healthcare initiative called the National Disability Insurance Scheme (NDIS).[2] The NDIS will fundamentally change the way in which:
>
> → organisations provide services
> → people receive services (participants)
> → people who provide services (carers) interact.
>
> Service providers originally received bulk funding from the government, which they could allocate to run their business. This has now changed. The government pays for services based on submissions the service provider makes to the NDIS, which has caused the business operating model to change. Carers now need to account much more closely for the time they spend with the people they are caring for. They also need to submit their time in an accurate manner so that claims can be made for the release of funds for the services provided.

Strategy

A strategy to tie your organisational goals to your people is crucial. When you develop a people strategy, the benefits you achieve from WFM, along with the benefits you achieve from HR, need to be tied together. Without this strategy, you do not have a complete view of the people in your organisation. Organisations that consciously look at their people in a holistic manner have a greater chance of achieving better business value. This also leads to setting realistic expectations with your people in terms of benefits they will receive.

WFM is at the core of these changes as it aligns what people do, when they do it and how long they take to do it. Using the challenges and benefits framework (see an example of this in Figure 7.2), you can quickly distil these business challenges into workforce benefits.

Once you understand the challenges and benefits, you can confirm them with senior management prior to embarking on any workforce-related initiatives. This is also a great tool to manage scope further down the track. Completing the challenges and benefits framework enables you to transfer many of these items to the benefit measurement framework defined in Step 5 - Measure.

Figure 7.2: Challenges and benefits framework in a healthcare NDIS example

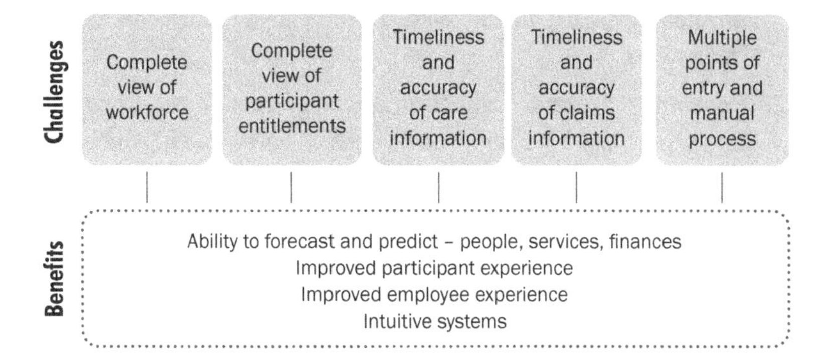

Roadmap

As we have seen, WFM provides many benefits but there are several factors to consider in the approach to achieving them. The outcome of these considerations is a roadmap that examines the key areas such as benefits, people, process, technology, priority, timeline and budget. The roadmap will enable you to align the benefits with your workforce and execute the strategy.

The roadmap takes into consideration tips for smarter implementation to ensure success in any WFM initiative.

We'll now look at some areas for consideration that may be relevant to your organisation. I have focused on these areas as they can have significant positive or negative influences on achieving your business goals.

WFM organisational maturity

Different organisations are at various stages of WFM organisational maturity. I have seen organisations embrace people-related change at a rapid pace and adopt many of the WFM benefits as part of a single

WFM initiative. The order of WFM adoption is usually completed as follows:

1. **Time recording** – electronically capturing start and stop times.
2. **Award interpretation** – feeding start and stop times to an electronic calculation engine that calculates the correct EBA payment rules, e.g. normal time, overtime, allowances and deductions.
3. **Rostering** – allocation of shifts and work on specific dates and times to people considering items such as coverage requirements, availability and skills required to complete this work.
4. **Forecasting/work-based/route** – applying additional rigour to the creation of a roster and allocation of work considering things such as sales volumes, patient intake, order of service to better determine the people requirements to meet demand, etc.
5. **Optimisation** – looking for further areas of optimisation within your workforce, e.g. the optimal mix of casual and full-time staff.

When you adopt all the previously defined WFM areas together, this results in your operational managers taking on additional people responsibility (as noted earlier in this chapter under the sub-heading Leadership). Your managers now have full visibility of their team and are delegated to make real-time people decisions. This step may be too great for your organisation to complete in one go, or your organisation may not want to give this responsibility to its managers.

Some organisations that have strong, established communication lines from senior management to the operational coal face are well positioned to complete holistic WFM adoption. The impact of change can be communicated, understood and adopted rapidly. In other words, if your senior management is committed to mass change and can get their people aligned to adopt the change, you can take on more areas of WFM adoption together.

Be sure you understand where your organisation's maturity sits to make an informed decision.

Multi-faceted organisations

Some organisations are made up of a collection of businesses and are therefore multi-faceted. For example:

→ a parent company may own three retail chains and a manufacturing organisation; or

→ a council may be responsible for a contact centre, works management, health services and an entertainment facility.

Multi-faceted organisations may not find a one-size-fits-all WFM process and/or solution. Each business and business area has its own unique requirements. We'll look at this in more detail in Step 2 – Prepare: What you need to know before you start a WFM initiative.

In the council example above, the four business areas (contact centre, works management, health services and an entertainment facility) may all require specific outcomes, which means each business area may need to have its own implementation initiatives. From an implementation perspective, you may find you have a number of implementations under a single program of work. You will also need different skillsets from each of these business areas to represent the current and end-state positions.

Departmental considerations

Today's WFM solutions have impacts across many areas of organisations. It is important to engage your departments to ensure the go-forward roadmap is adequately represented. Refer to Chapter 2, What does WFM mean for your people? and Chapter 3, What does WFM mean for your customers? for more on this.

Some practical considerations include:

→ Are your departments available to assist? For example: is the payroll department engaged in year-end activities? Is HR aligned to balance the WFM needs in the overall HR strategy?

→ Are the requirements of one department overarching another? For example: will the WFM initiative take priority over a new manufacturing initiative?

→ Are any incumbent systems approaching the end of their life? Do the systems need to be replaced?

→ Have your business needs surpassed your current systems functionality or the vendors capability to support you?

Prioritisation

When completing the roadmap, ensure that the organisational considerations are adequately represented and balanced. The roadmap will prove a valuable asset when moving into subsequent steps of your WFM journey.

As noted earlier, it is likely your organisation will undertake a multi-stage journey to reach an end state. I have successfully seen these roadmaps completed with current-state, interim-state and future-state operating scenarios.

Business case

The business case is a key outcome of the alignment step. It complements the roadmap and often changes, or the impacts of one will affect the other. The more effort you put into the business case, the more realistic the changes and costs will become to your organisation.

Implementation Planning Study

Some organisations complete an Implementation Planning Study (IPS) at this point in the project. The objective of the IPS is to take a deeper dive into your organisation to better understand how it operates, which will better inform the business case and provide further details for the roadmap. An IPS is analogous to a high-level planning and design phase for a project. It will generally look at potentially risky areas.

In my experience, an IPS is invaluable for multi-faceted organisations and organisations with autonomous operating models.

Details of the IPS are further fleshed out in the next step of the methodology: Step 2 – Prepare: What you need to know before you start a WFM initiative.

People, process and technology eco system

Items from a people, process and technology perspective all contribute to the business case, and we will look at each of these below. With the transition from on-premise to cloud, the considerations have changed from a people, process and technology perspective. In the past, technology drove outcomes. Today, people and process drive the outcomes, along with more careful consideration of senior management-stated business objectives. The key changes are depicted in Figure 7.3.

Figure 7.3: WFM eco systems past and future

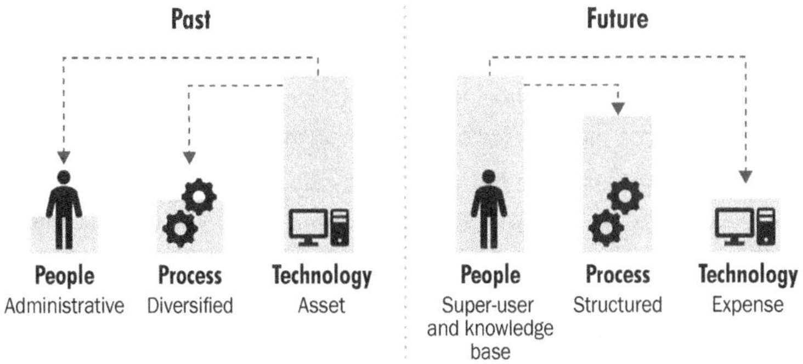

Past

People
Administrative

Process
Diversified

Technology
Asset

Future

People
Super-user and knowledge base

Process
Structured

Technology
Expense

People

People are at the centre of WFM initiatives and organisations place far more focus on people and positive experience today. This places a high emphasis on involving people heavily in your initiative, along with giving them responsibility to make decisions with appropriate support.

When you are in implementation mode, input from multiple perspectives is required. Consider the importance of 'backfill' with the critical roles on the project. In other words, when you take a person who works for your organisation and dedicate them to the WFM initiative, you will need to find a suitable replacement for that person's role. If not, you will be asking them to do their current job and whatever is expected from the WFM initiative. This can result in poor quality delivery due to lack of time or burnout.

Be pragmatic and ensure you budget for this adequately when you are calculating your implementation costs. If you don't get this right, it can lead to frustration among your team members, and a less optimal solution.

Once you have deployed the solution, consider the support required to operate the solution. For example, introducing the concept of a super user who can gain high knowledge of the people, process and technology required to support the organisation. Also take other considerations into play such as how many rostering staff will you require? And will you insource or outsource support?

Process

With an increased focus on improved people experience comes an increased need to provide localised processes. I'm not advocating that you build a process for every nuance in your organisation; rather, that you need to ensure you understand the way of operating and get the process right. To obtain the business benefits you should follow a defined process and the cost of implementing the new process must be considered. Your organisation's WFM maturity will drive the cost of this change. At the outset of your initiative, you may think a single process is possible and, after obtaining more knowledge about your business, you might require more than one process.

As an example, in the past I have seen some organisations implement, say, three processes for rostering and the specific business areas are able to choose the most appropriate process for them. For example:

- → Process 1: Base level - No roster costing
- → Process 2: Intermediate level - Roster costing by hours only
- → Process 3: Advanced level - Fully costed rosters.

Technology

We have moved from technology being treated as an asset to an expense. Financial models are now based around regular incremental payments. Ensure you understand how many technology solutions will be required to deliver your end state and how these technologies will coexist.

Consider technology that provides all your functions from a single application, or a technology layer that allows the user interactions to coexist across multiple technologies in a frictionless manner.

How many business cases?

If your organisation has a single operational model, it is likely a single business case will be put forward for the program of work.

Some larger multi-faceted businesses have multiple businesses within a business. Other businesses provide the same core functions to their clients but have different operating models across their locations. As people and people processes are at the heart of the businesses, different ways of operating can mean a few things or a combination thereof, for example:

- → the business will continue to have different operating models
- → the move to a standard operating model will require substantial change and adoption activities
- → a number of operating models will be adopted to move forward.

Where there are different operating models, some organisations break their WFM initiatives into multiple business cases. While on the surface this may appear to prolong the timeline, there are benefits:

- → each business case stands on its own merits
- → there are fewer initial costs due to multiple projects starting at different stages
- → you can get runs on the board quickly
- → organisational change and adoption is spread over a period of time
- → learnings from previous initiatives can be fed to future initiatives
- → a repeatable process can be developed for future initiatives
- → benefits and measures can be better understood each time around.

This approach can be particularly relevant for larger multi-faceted organisations. It comes back to the fundamentals: understand the change, culture and mindset shift required to transform your organisation and ... don't bite off more than you can chew.

Further considerations for Step 1 – Align

Let's take a look now in a little more detail at further considerations that are important in the align stage of WFM implementation.

Cost savings versus current practice

Organisations often embark on a WFM initiative with the expectation there will be cost savings as a direct consequence of the initiatives undertaken.

In some organisations, when EBAs are closely scrutinised or solution testing commences, it is occasionally found that underpayments have been occurring for some time. In extreme circumstances I have seen WFM initiatives concluded without even starting due to this scenario. The organisation decides there is less change and risk associated with leaving things as is. In Step 2 - Prepare: What you need to know before you start a WFM initiative I provide some tips for dealing with organisational compliance issues, such as underpayments and overpayments that need to be resolved in the course of the WFM project.

Off-the-shelf versus bespoke solutions

The market has reached a level of maturity where it is becoming less likely that an organisation will require a custom-written bespoke solution. When choosing an off-the-shelf solution, traditional considerations come into play when you are selecting the right supplier to best meet your WFM needs. Some suppliers are better aligned technologically to one industry over another. Make sure you speak with other customers in the same industry with like requirements to help you mitigate any technology risk in this area. That is, be careful not to select a technology solution that does not meet your needs.

The importance of leadership

The core of why I wrote this book and the basis for the 5-step methodology is that WFM projects are people projects, completed by the people for the people. With this in mind, to achieve the best results you need to own it and drive it.

In the past, businesses have traditionally focused on Step 3 – Implement: What you need to know to adopt WFM. This is also the key area product suppliers focus on, as often everything else is outside their core remit. This is even more prevalent these days as most product suppliers have venture capital driving their business growth and licence sales drive their valuation, i.e. implement with a light touch and move onto the next product sale. This is not a negative, it's just a commercial reality. It's important to recognise this and deal with it when you are setting up your WFM initiative because from a leadership perspective, you also need to focus on Steps 1, 2, 4 and 5 in the methodology.

Pilots

I recently participated in a pilot for a large global retailer where the pilot was conducted over a very short period. There was high senior management and end-user acceptance due to a strong alignment of organisational goals and workforce goals. This resulted in an accelerated sales and implementation process benefitting both the customer and vendor.

Although many organisations conduct pilots, most pilots fall short of meeting the organisation's expectations. Generally, this is a result of any or all of the following factors:

- → Organisational goals are not aligned with workforce goals, which results in a solution that does not solve the business pain.
- → Business users are expecting a complete (end-state) solution but the product supplier only builds a partial solution to save time and money. This results in the need for many manual processes and workarounds that deflate organisational/user expectations.

- → A small sample of 'easy' business areas is considered, but not the areas where the real pain is being felt.
- → Product suppliers do not push for this, as their preference is to 'sell' the complete solution from the outset without a pilot (which enables the product supplier to move onto the next product sales opportunity).
- → It is too hard to deploy the technology, across clients, servers, mobile solutions and connectivity.

With the advent of true cloud solutions, coupled with product suppliers building mobile-first solutions, it is now much quicker to deploy technology solutions. Many end-users within organisations, especially in the small-to-medium business (SMB) space may have subscribed to a WFM software and be using this already.

Impacts of supplier-provided templates

It is common for suppliers to provide templates for their customers to complete, because it allows the supplier to accelerate the configuration build required for an implementation. Types of templates might be used to capture people master data, locations of work, cost centre information, organisational roles and staff skills.

While this approach may keep the supplier cost down, it does not absolve you from the responsibility of completing the templates. In some respects, it transfers risk from the supplier to you; for example, risk of project delays due to non-completion, or the implications of making changes to the data in your templates. Make sure you understand the expectation the supplier has of you to complete these templates, and the outcomes that can be expected. There is more on templates in the section on Accelerators in the next chapter.

Process and role changes to achieve the benefits of WFM

A defined process must be followed to achieve the benefits of WFM. For example, if your organisational goal is to decrease operating costs, this may cascade to a workforce goal of a reduction in the amount of overtime. Your current process involves various overtime approvals; input from operational managers in some business areas, input of overtime by administration staff in other areas. To deliver this benefit, you want to make your operational managers more commercially aware, remove the need for administration staff to input overtime and ensure the operational manager has an increased visibility of overtime along with giving them the skills to manage this effectively. In this case, the business process and role will need to change to achieve the outcome.

Another example of process change might be where your organisational goal is to give your people a better working experience. This may cascade to a workforce goal of enabling flexibility for your people to set their availability to work on weekends. The current process involves all staff being required to work on weekends on a rotational basis. Again, the process needs to change to achieve this outcome. In Step 2 - Prepare: What you need to know before you start a WFM initiative we look at understanding the current state and preparation for change.

Duration of a WFM project

There is no easy way to gauge how long a WFM project will take to implement, but here are a couple of scenarios at each end of the spectrum to help put project duration in context.

If an organisation is ready for change, has a culture of change and understands its WFM maturity, it can deploy rapidly and on-scale. I have seen successful global deployments across multiple geographies and sites take place in a number of weeks.

For many organisations, however, timeline is not the key driver; rather, they focus on a successful outcome with low risk. Some organisations break the project into chunks by business area, complete a business case based on business area, and deploy based on business area. These initiatives can take several years to complete, but offer great benefits in terms of people experience and risk mitigation.

Cloud versus on-premise focus

One of the biggest benefits of cloud is the time allocation of your project manager (PM). Traditionally, let's say the PM spent one third of their time on people, process and technology respectively. With cloud, the technology activities take significantly less time, so this allows the excess to be reallocated to high value-adding business activities. The depiction of time allocation across on-premise and cloud for the project manager can be seen in Figure 7.4. A more typical time allocation may be 45 per cent people, 45 per cent process and 10 per cent technology. The additional time focused on people and process can be allocated to benefits such as improving employee experience, personalising the design and concentrating on business outcomes.

Figure 7.4: Project manager time allocation – on-premise vs cloud

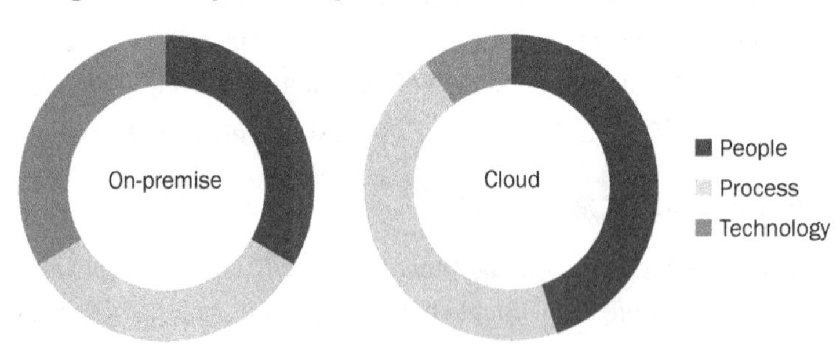

Impact of industry type

Industry type has a significant impact on WFM initiatives. The major impact is the method of rostering and how this method affects your people's way of working. This also has flow-on effects to requirements and the product and service suppliers who are suitable to your industry. These areas are discussed in more detail in the next chapter, but here are some examples:

→ **Manufacturing:** You require a fixed number of process workers on the morning shift, which runs from 7:00 am to 4:00 pm, five days per week. Shifts are fixed and repetitive.

- **Retail:** You are concerned about making sure you have your team members balanced to match the flow of customers into the store. Over the lunch period, customer numbers increase and additional staff are required. You require a pool of skilled casuals you can tap into quickly if team members call in sick. You require specific skills to greet the customer and personalise their experience. You require additional skills to complete the sale and gift-wrap the purchases.
- **Airline:** You need to consider which port the flight starts from, port closures, delays to flights and so on. This makes rostering highly dynamic and highly specialised due to minute-by-minute changes.

Industries where we traditionally saw WFM solutions are where there is a high percentage of workers who are paid on an hourly basis. Representative industries where WFM has a high fit include (grouped by rostering type which will be defined in the next chapter):

- manufacturing, public sector, transportation, mining, hospitality, education
- retail, health, contact centre
- construction, service
- home health, security.

We are also seeing the emergence of WFM in white-collar industries. The reasons for this might include:

- people wanting to set their availability so they can spend time with their children;
- the ability to handle contingent workforces; and
- the ability to work remotely and in diverse locations.

WFM – a journey or a destination?

The needs of your people and customers change regularly. Continual advancements in technology coupled with organisational maturity, all have an impact on your ability to adopt and change due to WFM initiatives. WFM is an ongoing journey; you need to continually measure its effectiveness and adjusting to change can harness great benefits for your organisation in the long term.

Also refer to Chapter 1, The evolution of workforce management, which demonstrates that the entire lifecycle is ever-evolving.

Top take outs

→ Align senior management to articulate its organisational goals and be a continual face to drive adoption activities.

→ Baseline your required business outcomes and continually refer back to them.

→ Create a strategy and roadmap to tie your organisational goals to your people goals.

→ The business case ties the overall initiative together. One or many business cases may be required.

→ Understand the journey your organisation is about to embark on from a people, process and technology perspective.

→ Industry knowledge is important to successfully deliver your outcomes.

→ Blue and white collar worker types all benefit from WFM.

Where to next?

Step 1 – Align focused on setting the foundation in place to achieve value from your WFM initiative. We'll now take a closer look at what is required in your organisation before you commence your initiative to ensure its success with Step 2 – Prepare.

---

1 E. Garton, 2017, 'What If Companies Managed People as Carefully as They Manage Money?', *Harvard Business Review*, 24 May, viewed 12 July 2017, <https://hbr.org/product/what-if-companies-managed-people-as-carefully-as-they-manage-money/H03ODJ-PDF-ENG?referral=03069>.

2 NDIS, Home Page, viewed 5 August 2017, <https://www.ndis.gov.au>.

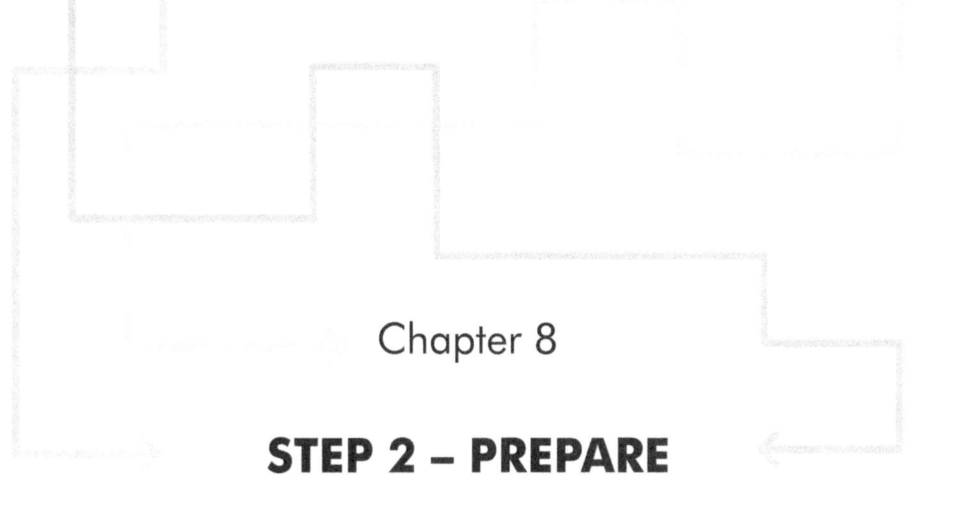

Chapter 8

STEP 2 – PREPARE

Too often I see a disconnect between the business outcomes senior management is looking to achieve versus what gets implemented. In extreme cases large financial investments are made, with little WFM solution actually being implemented. From my perspective, the biggest reason for this is that the organisation does not know what it needs to do prior to an implementation commencing. The consultants turn up to start an implementation with a statement of work that has some form of loose coupling to the business case. The consultants are ready to implement the technology and are well versed in the questions they need answered to configure the product. The organisation, on the other hand, has many unanswered questions prior to having a discussion with an implementation consultant or completing a product configuration questionnaire.

What you need to know before you start a WFM initiative

Step 2 – Prepare focuses on ensuring your business is educated and ready for the WFM transformation journey ahead (see Figure 8.1). This includes understanding:

→ your requirements
→ what type of rostering models are needed

→ which suppliers are suited to particular industries
→ how to contract with suppliers
→ the impacts of change
→ how your departments align
→ what business impacts will take place.

This chapter will unpack these key considerations to enable your implementation to be a success.

Figure 8.1: Smart WFM Methodology: Step 2 – Prepare

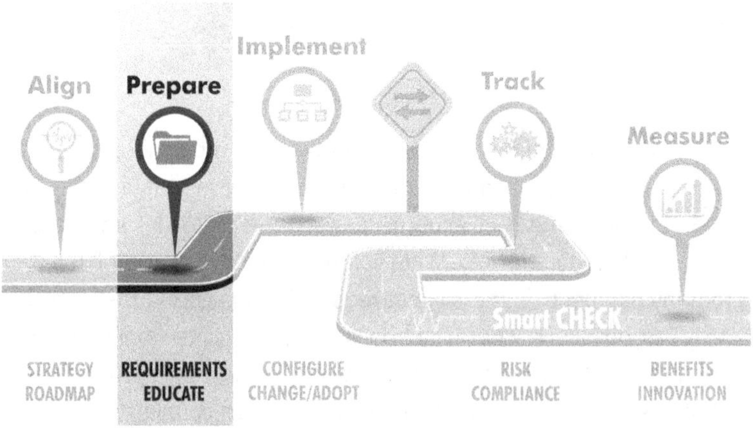

Understanding rostering models

Getting the rostering right in any organisation is key to business success. The more effectively you can manage your workforce to cover the work to be performed, the higher your returns will be. WFM has the ability to transform your business, providing you know what you want to achieve, in other words you need to be prepared.

Rostering or scheduling?

In the Smart WFM Methodology there is no difference between rostering and scheduling – the terms can be used interchangeably.

In some organisations or across different suppliers, these terms mean different things. The term rostering can be used to allocate workers over a block of time, for example a day. Scheduling can be used to secure a worker to carry out a discrete activity that is completed during the day, for example, completing a service call.

Types of rostering

Before you embark on any improvement initiatives, it is important you understand the method of rostering has a direct correlation to the way your operational workforce will function. This helps to frame the changes that will be required to deliver the business benefits. In my experience, rostering in organisations generally falls into four categories, which are described below. All industries start with shift-based rostering and some industries will build upon this baseline.

1. Shift-based

Shift-based rostering is the starting point for any organisation that rosters. There is a start time and a stop time for a shift. Often the shift will fall into a cyclic pattern, but it could also be an ad-hoc, once-off shift.

> **Example:** Person 1: 8:00–4:00 pm, Monday to Friday. Repeated weekly. This is analogous to setting up a calendar entry in your favourite email system.

Shift-based rostering will allow you to achieve the following business functions:

- → **Determine people's availability** - i.e. an employee can set a preference for availability to work on a given day/time.
- → **Align skills to a shift** - e.g. you need one person on the shift to have a first aid certificate.
- → **The ability to complete shift swaps** - e.g. a person can swap their shift with another person who is available to work and has the required skills.
- → **The ability to capture start and stop times of shift** -e.g. a person starts and stops a shift on a given date and time.
- → **The ability to cost the time on the shift** - e.g. to determine which cost centre the shift cost should be allocated to.

Typical industries that use shift-based rostering include manufacturing, public sector, transportation, mining, hospitality and education.

2. Forecasting

Forecasting takes in additional factors to forecast labour demand for the future. These factors enable you to better align your people to the future demand on a particular day or time.

Example:

- → In retail, forecasting allows you to align historical sales value by department to the number of people required to make the sales. Foot traffic is aligned to determine the number of people required to staff the store.
- → In health, forecasting allows you to align patient admission by department to the number of people required to treat patients.
- → In contact centres, you can align call volume with the number of people required to take the calls. For example, a product recall in a retail store relates to a predicted call volume.

Also refer to the section on labour standards below, which provides an overview of the process required to determine how long activities take, to help work out the optimal number of staff to roster on.

3. Work-based

Work-based rostering is based on the completion of jobs or projects. People are often allocated to teams/gangs/crews and complete multiple tasks to undertake a job or project at a specific location. These jobs or projects take place over variable periods (hours, days, weeks or months).

Example:

- → Team 1. Available 8:00–4:00 pm, Monday to Friday.
  Repeated weekly
- → Job 1. Monday: Fix damaged road.
  Tasks: Pothole repair, fill cracks, seal coat.

Typical industries that use work-based rostering are construction and repairs and maintenance.

4. Route

Route rostering is based on the servicing of clients at given locations. People often work individually or in small teams and complete the same tasks for the same client repeatedly.

Example:

- → Service Provider 1 available 8:00–4:00 pm, Monday to Friday. Repeated weekly.
- → Client 1. Monday: Domestic assistance. Tasks: Prepare meal, shower, groom.

Typical industries that use route-rostering are home health and security.

Labour standards

Labour standards are a way to baseline the average time taken to complete a discrete piece of work. For example, the time it takes to serve a customer at a supermarket checkout.

To develop labour standards, detailed time-in-motion studies are completed to determine the overall time taken to complete a task or a group of tasks.

Labour standards are prevalent for the forecasting, work and route-based rostering models previously defined. The benefit of understanding labour standards is that based on the standard, you can better align and justify the required number of staff to the demand.

Example: A retail store is expecting 450 sales over the 60-minute lunch period. One person can serve 45 customers in an hour. The standard clearly shows that multiple staff are required to service the expected demand. The exact number of staff will be determined based on the average customer wait time the store is willing to accept, and any other factors the store considers relevant.

Roster optimisation

While labour standards help you align staffing to demand, roster optimisation helps you determine the optimal roster configuration to meet business rules. Roster optimisation may look like this:

1. Alignment of labour standards to required staffing.
2. Meeting obligations as defined in the EBA, for example break between shifts, staffing requirements (senior and junior to be rostered together).
3. Meeting business objectives, for example two seniors preferred to be rostered during the lunch period.

Roster optimisation is typically a complex analytical activity and requires specialist skills to complete. Not all product suppliers' software will enable this. These tasks are often completed outside the WFM technology layer and the results are fed back into the technology.

Some organisations choose to implement all these items together, while others do it singularly or use a combination thereof.

Centralised versus decentralised rostering

Centralised rostering is completed from a hub where all your roster team works for the benefit of the many locations within your organisation. Decentralised rostering is completed at specific locations within your organisation. Whether you complete centralised or decentralised rostering you will require people with specialised skills in your roster team to undertake this activity to drive the optimal results. There is no right or wrong answer when making deciding upon a centralised or decentralised rostering approach, it's a matter of what works best for your organisation.

Here are some considerations when choosing what rostering type is right for your business.

Risk

Some businesses have functions and roles that are critical to successful operations. For example, in an airport you must have a luggage screener so passengers can clear security to hop on the plane and there are commercial obligations you must meet. When you have a critical need, centralised rostering models can be attractive. You are able to ensure that all critical screening roles are filled across all your locations and you have full control over making quick changes if roles are not filled.

Skills

Decentralised rostering models require specialised skills to be spread across all your rostering locations to complete rostering. If you have a business where there is a high turnover of staff, you will have to continually educate new staff. This means you will have additional costs of education to cover, and you'll need to consider which methods of education are best, and so on.

Personalisation

If you have a well-skilled rostering team, you can personalise the rostering experience for your team. In decentralised models this is particularly relevant as the rostering staff will develop a personal relationship with their people, often leading to improved employee experience. On the flip side, if the quality of the rostering staff is not high, it can have the opposite effect on your employee experience.

Data

When you have a highly skilled rostering team, the accuracy of your rosters will be high. The higher the accuracy of the roster, the higher the quality of data. Once you start reporting and analysing the rosters, you will be able to report more accurately on such things as the number of roster changes compared to forecast, variance of roster to actual time worked, etc.

Alignment of requirements to goals and benefits

It's important to first define the workforce goals that overarch the high-level requirements. Requirements are more functional in nature and are often defined in a singular manner. For example, they may include the:

→ ability to create a roster

- → ability to create a roster pattern
- → ability to cost a roster.

To understand what you are looking to achieve from a goals and benefits perspective, it helps to understand what high-level requirements are important to you. You also need to keep the end-user firmly in mind at this stage in the process.

Some WFM suppliers are able to provide a baseline set of requirements that you can take as a starting point. Your industry type is important when understanding your requirements, and specific requirements will be relevant to particular industries. For example, some industries will need the ability to:

- → allocate crews to jobs to fix a road in a service industry; or
- → roster care workers to clients, enabling domestic assistance activities to be completed in home health care.

Reporting and analytic considerations

Reporting and analytic considerations vary considerably. Here are my observations on some of the main ones, along with some practical examples.

Operational and management reporting

Operational and management reporting enables user-based analysis, generally around day of operations or within a pay/roster period. Typical examples in WFM are:

- → actual cost versus schedule cost reporting (such as normal hours, overtime hours)
- → roster reporting (such as who is rostered to what shift, location of shift)
- → effective roster reporting (percentage of time the roster meets forecast).

Metrics come into play with this type of reporting. For example, when production line labour cost approaches 95 per cent of budget, the operational manager should receive a notification.

WFM reporting is different to payroll or cost-accounting reporting in that it is available in real time. This allows a manager to make an informed decision to proactively address a scenario before or as it is occurring. The old saying comes into play: prevention is better than cure.

> **For example:** The labour cost planned for the day shift tomorrow is $X and is running at 95 per cent of budget. Changes can be made prior to the shift to reduce the cost of the shift.

Predictive and prescriptive analytics

With the growing use of AI, machine learning and deep learning, we are seeing an increase in the value of predictive and prescriptive analytics. Predictive analytics use data to predict the likelihood of a scenario occurring. For example, a person might have a history of taking a day off after they have worked a specific shift combination. To remedy this, the system may prescriptively suggest a change be made to the people making up specific shifts.

Specific data science roles that help identify and resolve trends that are occurring within organisations are becoming more prevalent.

Data

Data cleanliness is imperative to enable decisions to be made based on what has really happened or is happening. The earlier example notifies management if labour cost approaches 95 per cent of budget. If the data is inaccurate, the cost might be at 105 per cent of budget when the decision is made. Driving a culture within your team to ensure your data is accurate and 'clean' is crucial to making correct decisions. Also refer to the section on data cleanliness in the following chapter where I address this from an implementation perspective.

Suppliers

It is common to use a combination of product and service suppliers to provide your overall solution. However, there are several important considerations that apply to selecting the right product or service suppliers for you.

When evaluating suppliers, don't just pay attention to *singular* business requirements as defined in a document such as a request for proposal (RFP) or a request for information (RFI). Look at requirements collectively to ensure that ALL your collective business requirements can be met. This is also be known as a 'use case'. Taking a design-thinking and persona/process approach to this task can be beneficial. Consider a frictionless experience from an end-user perspective.

Industry considerations

Consider your supplier's fit to the industry that you work in; the terminology you use, the way you work, market trends and so on. Is the supplier aligned to your goals, benefits and requirements? Can the supplier play out and understand your business persona? For example, in the health industry, what does WFM mean for a nurse, or for a nursing unit manager?

Technology

Consider your supplier's technology alignment to the industry you work in. Different suppliers work more effectively in different industries. For example: one supplier may have a strong fit to retail but a low fit to health. Another supplier may have a strong fit to clinical health but a low fit to home-based health care.

Small-to-medium business versus enterprise business

Some suppliers work better with SMBs, while others are a better fit for enterprise business. Let's define the industry sizes as follows:

- **Small** < 500 employees
- **Medium** > 501 < 2,500 employees
- **Enterprise** 2,501+ employees

These numbers are not absolute and should not be taken literally. Some SMBs aspire to become enterprise and they potentially should be treated as enterprise at the time of making a WFM supplier selection.

Take the time to understand where your supplier's experience lies. In enterprise you may expect, for example, provision of strong governance, account management and on-site resourcing, to name just a few requirements. If you are choosing an SMB supplier to provide an enterprise solution, make sure they can meet your expectations and/or have a suitable partner to assist.

Accelerators

Many suppliers will offer accelerators to facilitate your WFM initiative. Accelerators generally fall into the following three categories:

1. **Tools:** Methodologies, templates, scorecards used to inform your WFM initiative. This book is an example of a tool.
2. **Business led:** Used to capture information required by the product supplier to build the product. The supplier will expect you to complete templates, generally with minimal guidance. The templates are a way for product suppliers to keep their costs down, but they don't always inform you of the implications of what providing or not providing the data means for your business and their software. For example, your data informs the design of your rostering model. The adopted design does not cater for your business to grow into additional geographies. The implications to fix this require significant reconfiguration and data realignment. This is an extreme example, but nonetheless a scenario that could occur.
3. **Technology:** Used as a pre-configuration or baseline to accelerate your configuration. The product supplier will expect you to adopt these pre-configurations as provided. Again, these pre-configurations are a way for product suppliers to keep their costs down, but they don't always suit your business rules.

Understand what the supplier expectation is regarding variances and associated commercial implications if the accelerators are not suited to your business.

We will look further into accelerators and what they mean to your implementation from a business and technology perspective in the next chapter.

Roadmap and support

Understanding the supplier's roadmap and investment in R&D will help you make your decision choosing the right supplier. If the supplier can show you a journey of where their organisation is heading, you will have more confidence in their ability to be with you for the long term.

It is important to understand items including:

→ **analytics capability:** especially correlation and prediction

→ **AI:** particularly machine learning and automation

→ **user experience:** Is the process frictionless?

→ **omni channel:** Is the platform capable of taking and processing multi-channel data in a seamless way, e.g. from social media, voice, IoT?

→ **cloud:** Is the product 100% cloud-based and built on a single data model?

The ability of the supplier and/or the partner ecosystem to support the WFM solution is another consideration. Once you are live, you need confidence that any issues that arise will be dealt with promptly. WFM solutions are operational in nature and any downtime can have significant business impacts.

Cultural fit

I am a firm believer that a cultural fit is critical for the overall success of a WFM initiative. Is the supplier just trying to sell you a product or service, or are they really trying to understand you and your business? Does the supplier want to undertake a journey with you?

The impacts of WFM across your business are far-reaching, and if you can't align on a cultural/emotional level, it places a question mark over the ability for your supplier to work with you to undertake your organisational transformation.

Will the supplier be honest with you - and will you be honest with them?

Technology

The next part in your preparation is to consider all things digital.

Digital strategy

Does your organisation have a digital strategy and are you aligned to it? Your IT department will have standards related to supplier credentials, security, availability, support, integration, architecture and so on.

In Chapter 7, I go into more detail about aligning technology.

Deployment methods

We currently have two primary methods of technology delivery: on-premise and cloud. In my view, cloud is further divided into 'on-premise in the cloud' and 'true cloud', which I discuss further in the following three sections: On-premise, On-premise in the cloud and True cloud.

A recent survey produced by AppsRunTheWorld[1] found that:

- in 2016 the percentage of cloud deployments in the HCM space was 64 per cent versus on-premise 36 per cent
- by contrast, in 2021 the percentage of cloud deployments in the HCM space will be 81 per cent versus on-premise 19 per cent.

This demonstrates the continued shift of technology to the cloud, which has great benefits for rapid deployment and adoption.

On-premise

WFM is a highly operational system by nature and it impacts those who work in your business on a day-to-day basis. The impact of on-premise, especially from the 1970s to the 2000s, was that the WFM technology had to be deployed to the client's server and desktop.

In a world of client server, on-premise made for very complex technology deployments. Project teams would often be spending their valuable time on technical issues, for example:

- → updating service packs;
- → attending to desktop compatibility issues;
- → deploying time-collection devices across diverse geographies; and
- → integrating bespoke systems to multiple coupled systems.

On-premise had a direct consequence on many areas associated with returning business value, for example it:

- → reduced the time spent working with business stakeholders to ensure successful adoption;
- → increased the number of issues and risks associated with successful technology deployment;
- → increased the total cost of ownership; and
- → improved overall time to value, i.e. the time it takes to receive the benefits brought about by adoption of cloud technology.

From a financial perspective, on-premise implementations are generally purchased outright, require yearly maintenance payment and additional services are pay-as-you-go.

On-premise in the cloud

Some technologies available today are still on-premise, but they are delivered in a cloud environment. Many of the challenges as described in the on-premise section are still relevant.

The main change is that the infrastructure is generally provided as a service (IaaS). While an organisation's IT department is no longer responsible for looking after the infrastructure, the service provider will still have the same challenges that the IT department faced.

From a financial perspective, these implementations are generally sold as PEPM and pay-as-you-go services.

True cloud

These solutions are 100 per cent cloud based. The impact of this, especially from the 2010s forward, is that the technology deployment is to a browser or mobile device. Many of the technical challenges that exist with on-premise are removed.

This brings different considerations, for example:

→ new functional releases are more regular, mostly with new functions 'shipped off';
→ it is not as common for the technology suppliers to make test environments available;
→ many of the technology suppliers are Millennials/Gen Z and lack enterprise client thinking; and
→ data exists 'off-site'.

True cloud provides many benefits including:

→ the ability to achieve time-to-value quickly. You can literally switch a solution on immediately;
→ the technology is 'mobile first'. Solutions are designed to work mobile first with web browser second; and
→ end-users can 'touch and feel' the solution, facilitating employee experience and adoption.

Some of the more mature cloud solutions have WFM and broader HCM functions in a single technology and on a single data model. This enables you to see in real time the impact of a roster change and what this will mean to a payroll, including on-costs.

Time capture

Today we have many options for time collection. Some of the common ones are described below.

Biometric

With biometric time-keeping, generally a fingerprint is used to authenticate identity. This method is popular in organisations where prevention of time theft is a high priority.

Note: No suppliers I have experienced store an actual fingerprint. A mathematical representation of the fingerprint is usually stored, which cannot be reverse-engineered to an actual fingerprint.

Facial recognition

Using a person's face to identify them can be implemented either by:

- → taking a photo of the person and storing it in the WFM database. If required, the actual photo can later be called upon to verify that person's identity; or
- → using the person's face to identify them.

The concept of this approach is similar to biometrics, with similar benefits. The main difference is that biometrics requires specific hardware (i.e. a finger scanner) whereas facial recognition can be implemented via software activation on most devices.

Mobility and geo-fencing

Mobile devices provide the ability to record information related to a shift. In addition, using a set of allowable GPS coordinates can define in what area a person can start and stop work using a mobile device. This is particularly useful for field service industries, and it enables employees to provide evidence of the location where and times when the work was performed. Other relevant information can also be captured such as job information, including materials used in field service industries.

Fixed devices

Fixed devices to record information related to a shift are beneficial when you want to ensure an employee's start and stop times are completed at the same place. These devices can be used for kiosk-type functionality such as costing transfers, applications for leave and viewing rosters. These devices are sometimes provided by the supplier or are available as commercial off-the-shelf devices (e.g. a tablet or iPad).

Wearables

Wearables are becoming common in the workplace and application of technologies such as RFID allow automatic capture of shift information. For more on wearables, refer to Chapter 5, The future: How will WFM deliver and interact?

Contracting

When working with a supplier, it is important to understand the different considerations related to on-premise and cloud contracting.

On-premise

With on-premise contracting, the supplier often tries to sell upfront the licences required now, plus those licences needed in the following years. To make this more enticing, the supplier might apply some form of discount. From a supplier's perspective, this model enables most of the revenue to be received upfront. Figure 8.2 shows how this may look over a five-year period.

Figure 8.2: Revenue spread on-premise

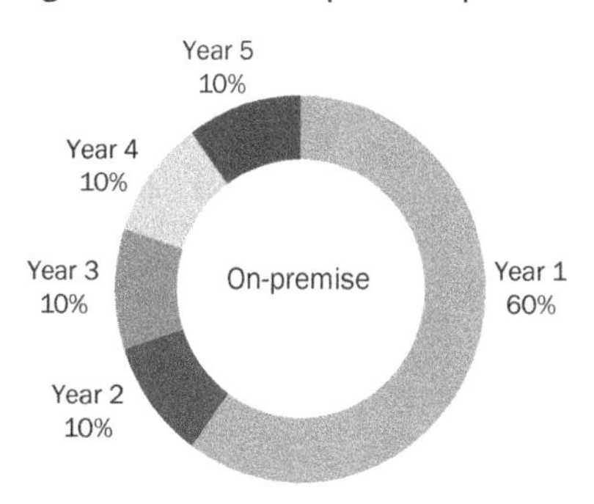

The implication from a business perspective is that once the initial sale is made, there is less reason for the vendor to stay accountable as their business client is locked in. It is common for licences to be purchased that were planned for future use. Many factors from a business and supplier perspective will determine if the licences are actually used or not.

> **For example:** An organisation may have purchased licences for roster forecasting with every intention of delivering these business outcomes at the start of the engagement. However, once the project starts, it is found that the organisational maturity is lower than expected and only shift-based rostering is possible.

Cloud

With cloud purchases, you have much greater control over the sales process given you only purchase what you need when you need it. In turn, this may only be for an initial pilot. From a supplier's perspective this model spreads the revenue over a greater period. Figure 8.3 shows how this looks over a five-year period.

From a business perspective, once the initial sale is made there is greater incentive for the vendor to stay accountable, as you may go elsewhere if you are not happy with the solution or service.

At this time, most cloud suppliers of WFM technology (especially SMB suppliers) have a 'light touch' implementation and support model. This enables the product supplier to focus on licence sales. Take the time to explore the service engagement model with your supplier to ensure you get the service you are expecting.

Figure 8.3: Revenue spread cloud

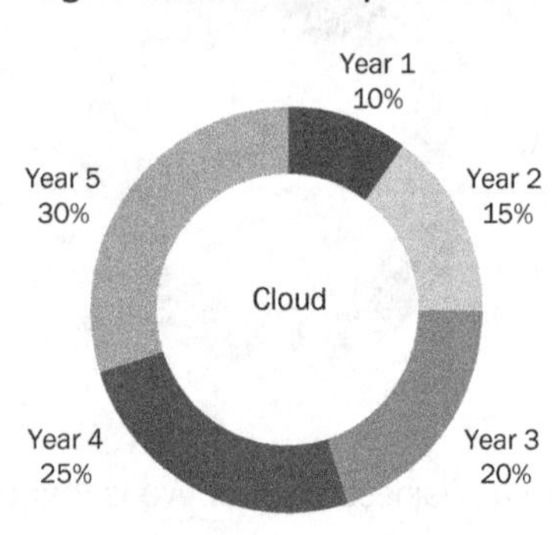

The contract

You generally have to negotiate three core agreements when you purchase a WFM solution:

1. Product.
2. Master Service Agreement (MSA).
3. Statement of Work (SOW) for the initial implementation.

Make sure you take the time to understand these agreements, what you are purchasing and your overall commitment.

With on-premise and on-premise in the cloud, most product suppliers generally look to contract for a specific number of years. The product supplier builds its financial modelling off the assumption it will be retained for a defined period. With on-premise in the cloud, the supplier's investment (or your investment) in infrastructure can be substantial. If you want to break these agreements, often you will have to pay a fee calculated on the number of months remaining in the term of the contract multiplied by the monthly PEPM fee.

Cloud agreements are usually more straightforward; there's usually an agreed term that can be terminated with an agreed notice period.

Understanding responsibilities is also a key area of contracting. Consider the inclusion of a Responsible, Accountable, Consulted, Informed (RACI) clause to ensure expectations are clearly understood from the inception of the project.

Buyer and seller involvement

Ensure the buyer and the seller are engaged throughout the life of the project or have handed over the project adequately. On occasions I have seen a solution sold with a well-defined business outcome. After contracting, however, the customer and supplier owners move on with little or no handover. This generally results in misalignment with the delivery team and business outcomes being compromised. The people and financial implications on your organisation can be dire.

Business impacts

We covered the impacts of WFM on your people and your customers in Chapters 2 and 3 respectively. This section explains key areas in preparation for a WFM initiative that require closer attention.

Award interpretation

Award interpretation is core to many WFM initiatives. This is the process of taking an EBA and determining what will be configured. In Australia, the Fair Work Commission[1] governs the agreements.

Can you configure directly from an EBA?

An EBA will define things such as rates of pay and employment conditions, e.g. hours of work, meal breaks, overtime. These areas are important to the overall interpretation of employee payments, but they do not define all the business rules, for example:

- If a shift runs over midnight, is the shift deemed to have started on the day the shift starts, the day the shift ends, or the day where the majority of the hours worked fall?
- If an employee starts work four minutes before the scheduled start of a shift, is the employee paid overtime, or is the start time rounded to the start of the shift?
- How do you handle words like 'should' that appear in an EBA? This word is not definitive in its nature and requires analysis of how it may be interpreted.

In my experience, the configuration of the award is not the time-consuming part. Most time is spent on the analysis and determining what will be configured. Do not underestimate the importance and the time required to successfully complete this step. It has a direct impact on the payments you will make to your people via payroll.

Some of the SMB solutions available on the market have pre-built EBAs. These generally have the baseline Modern Award configured. As a business owner, it is your responsibility to make sure you are comfortable with the interpretation and the payment of the EBA conditions.

Pay to EBA or site practice

When analysing your EBA, it is common to find that what is documented in the award is not what is being paid. In most situations when this arises, the site has made a localised interpretation of the EBA and pays to this. You will need to make a business decision whether to revert to the EBA or pay to site practice.

Sometimes senior leadership teams make a blanket statement that they intend to revert to an EBA. When the rubber hits the road, they find that current site practice is more generous and the risk of industrial action and change is too great. In these situations, the leadership teams will often leave payment as per site practice. Remember that if you pay to site practice, the cost to implement and support site-specific agreements may increase.

At the end of the day, this is a business call. Make the call at this stage of the project and give the delivery team clear guidance. Chopping and changing with these decisions can be costly.

Who makes the interpretation?

It is important to understand who in your organisation is able to determine how an award is interpreted. There is no one size fits all here, as some of the following examples show. In some organisations:

- → payroll makes the interpretation decisions based on timesheet data
- → operations make the payment decisions based on actual time worked
- → payroll makes an initial assessment and clarifications are made by operations or payroll.

Refer to Chapter 2, What does WFM mean for your people? - specifically in the section, WFM organisational impact, where Jim is unsure of his responsibility related to award interpretation.

Moving forward, it is important you determine who will be responsible for this now, as opposed to who does this in the future and who will make these calls on the implementation project.

Different interpretations of the same EBA

In many organisations the same EBA is in use across multiple sites and regions. It is common for interpretations to be different across the business areas. Simply put, the way one person interprets a clause may be different to the way another person interprets a clause.

Also consider grandfather clauses, i.e. clauses that are carved out for historical reasons applicable to one or many employees. These are often highly localised and known by only a few people in the organisation. Working through this to reach an agreed outcome can be time-consuming.

What to configure?

Many EBAs have conditions that are not paid under normal circumstances. For example, there could be conditions for a certain group of workers, but your organisation no longer has this group of workers. Will you configure these rules or not? A business decision needs to be made at the outset of a project to ensure that expectations are set correctly.

Ensure you agree with this as part of contracting with your service provider. The costs of configuring non-used areas of an EBA can be considerable for both you and your service provider. Remember: design, build, test, support.

Underpayment and overpayment

When reviewing EBAs it's not uncommon for the business analysis to identify under and overpayments. Establishing a treatment plan to address this early is recommended. Considerations may include back-dating and changes to the date of the last EBA. Work on the concept of netting over and underpayments.

Award interpretation key learning

Award interpretations impact project timelines and budgets more often than any other area. Generally, the larger the organisation and the more autonomous its operation, the higher the risks associated with this. Many organisations establish governance committees or working groups at the commencement of the WFM initiative to work through these considerations.

Business/union consultation

Consultation with the relevant union(s) from the outset is imperative in any WFM initiative. Being clear with union representatives on the changes that are coming and keeping them abreast of this with regular communication is key. In many organisations it is important to get input from your industrial relations (IR) department, that is a conduit between your people and your management.

People

Selecting the right sponsor is pivotal to ensuring overall success of the WFM initiative. If you want to transform your organisation, choose a sponsor that has the drive and influence in the organisation to make this happen.

People in a variety of business functions can take on the sponsor role:

→ operations

→ HR

→ finance.

In my experience, the greatest positive impact is when the role is adopted by operations and/or HR. People working in these areas have the farthest reach into the operational parts of the business, which is where the changes take place. The sponsor needs to have tenacity and drive to achieve the desired business outcomes.

Project manager

Arguably, this is the most vital role on the project. The role of the project manager (PM) will keep your ship pointed in the right direction every minute of every day. Make sure your PM understands WFM and can draw from real-life experiences to keep you informed, facilitate decision-making, set expectations and manage issues, risks and costs.

Engage the PM at the outset of the initiative and ideally get them involved with the contracting process. This will enable a smooth transition to deliver the desired business outcomes and to ensure they fully understand the scope of work.

Change management

More recently, some organisations have moved to use a combination of the project sponsor, PM, operations management and HR to collectively manage change. When undertaking a business transformation, credibility comes from those who are tightly coupled/aligned to the team members. When you set up a two-way communication stream between these roles it makes for very effective and clear messaging between the:

- → team member
- → operations manager
- → HR/PM
- → project sponsor.

I am not advocating that a specialised change management role is not required; rather, ensure you put the right process in place to effect successful change in your organisation.

There is more about change and the various roles and experiences in Chapter 9, Step 3 – Implement.

Project roles and skills

Selection of the right representative team on the project is instrumental to the overall success of your WFM initiatives. You are dealing with people, their emotions and their livelihood, so I would encourage you to have your brightest minds working on delivering the initiative. The key touchpoints are:

→ **Operational process:** The methods of operating will likely change with the introduction of WFM. For example, in a service industry, the way time is collected on completed jobs may be manual and in the future it will be via a tablet. In other organisations, there may be two business types: service and manufacturing. In order to understand your business operating models, you require representation from each of the operational areas to help shape the future way of operating.

→ **Financial understanding:** WFM initiatives are generally looking to improve the accuracy of costing to improve financial forecasting. It is important you have strong financial representation, particularly from a cost-accounting and reporting perspective.

→ **Payroll:** Times that are collected are award-interpreted producing hours, which are fed to payroll for payment, e.g. normal time, overtime. Ensuring your payroll team is involved will help reduce any risk associated with pay results.

→ **IR:** WFM projects will often identify pay anomalies or change the way of working. There can be clarifications and/or changes required related to the EBA. Having IR reps involved from the outset will help drive these clarifications and/or changes.

→ **Your people:** Your people will be impacted from a number of perspectives: the way they work, potential redundancies, efficiency improvements, repurposing, changes to job descriptions, downsizing and so on. Having HR and senior management involvement will help make the transition smooth.

→ **Your customers:** The impacts on your customers should be positive. Your staff should have less need to focus on administrative tasks and be able to concentrate more on personalisation.

Project role time requirements

There is no simple answer to the question of time requirements for each role. It depends on a number of factors such as the size of your organisation, number of staff available by business area, which roles take on what project function and so on.

The most important thing to realise is that many areas of your business are impacted by a WFM project and to achieve the best outcome people will need to work together. Freeing up the time of these specialists is often one of the greatest challenges to achieving your outcomes.

I've looked at the answer to this question of time requirement through a business and Smart WFM Methodology lens to help provide a flavour of what you might expect across the life of the initiative.

Figure 8.4 represents the time required across each of the business roles over the course of the WFM initiative. Note: your project will also have a core project team in addition to this. This diagram reflects how much effort you may need to expect of your business representatives. Enterprise WFM initiatives will often have full-time employees allocated to many of the roles.

Figure 8.4: Business role input

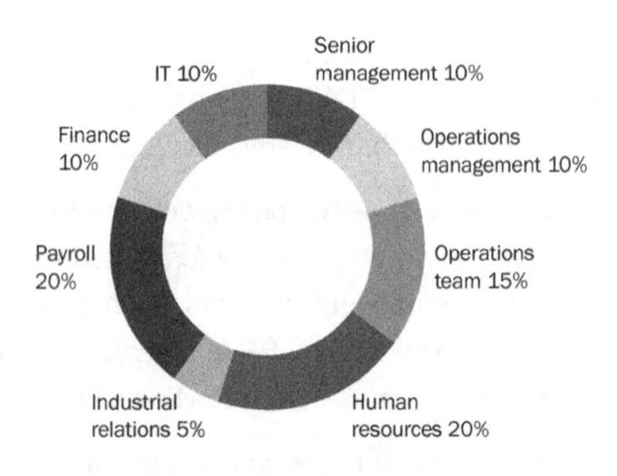

Figure 8.5 represents the approximate distribution of time required across each of the Smart WFM Methodology steps over the course of the WFM initiative.

Figure 8.5: Smart WFM Methodology

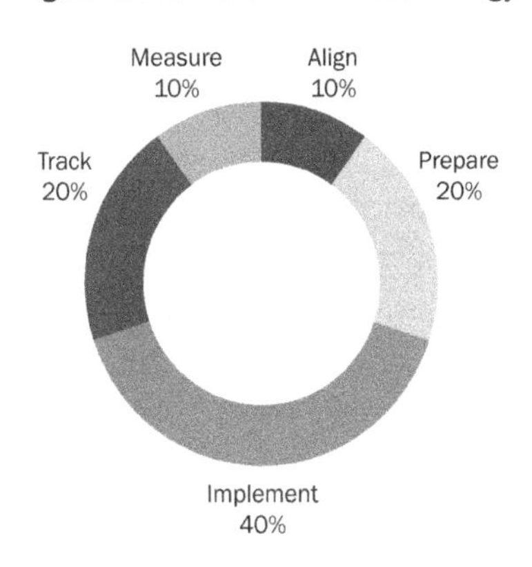

Business user role impacts

The greatest changes are seen in the operational area of your business, focused around operational and payroll improvement:

→ **Operations:** The impacts are seen from a tactical and strategic perspective. Tactically, operations managers are empowered to set rosters, complete time approvals and manage leave, to name a few affected areas. Strategically, operations managers have greater analytical visibility and understanding of what their team members are doing from a staffing and financial perspective. This enables them to become more commercially aware, develop their team and spend time on customer experience improvements.

   Having operations managers become more commercially aware has greater benefits in your organisation as this can lead to time being spent driving a greater sense of purpose and ownership for the team members. This then has a direct impact on the operations manager as they can develop skills that allow them to grow from a professional and personal perspective. Ultimately, this can result in your operations managers becoming more business-like in nature.

→ **Payroll:** WFM initiatives result in the automation of hours calculation for people's pay and the automation of hours transfer to payroll. This has a direct impact on payroll, allowing them to concentrate on value-adding activities.

People key learning

In many organisations, payroll completes a lion's share of the tactical activities that operations managers often undertake in a future state. These organisations find they have substantial change associated with the implementation of WFM. This is generally helping operational managers understand the importance and benefits associated with completing the tactical and strategic activities.

Implementation considerations

To wrap up this chapter on preparation, let's consider these final implementation considerations.

Global, regional and local influences

Larger organisations will work across countries, regions and locations. This will result in some practices that may need to be adopted and/or modified moving forward. For example, a particular location may have a public holiday on a defined day; some regions may have cultures that don't work well together; some countries might have differing business goals based on societal maturity.

Your ideal goal is to adopt global organisational standards where possible, but understand that regional and local influences will come into play. The more you localise, the more support you will need to give to the solution. Ask yourself two key questions:

1. Are we localising because there is a true business need to do so?
2. Are we able to redefine our business rules to look for standardisation here?

Recall the fundamental requirement to not bite off more than you can chew. On the surface, it may seem like a clever idea to standardise, but the reality to make this happen may be different. Be realistic.

Understanding the current state

Implementation of WFM will often require a change from one operating state to another. If you are going to achieve business benefits, you will be required to adopt a particular way of operating. Understanding the current state enables you to understand how great the change is and to inform the design. The next step - Implement - begins with the concept of a survey to help you understand the current state.

It is important to understand the current state prior to commencing implementation. Many WFM initiatives stall or don't deliver the desired business benefits when the current state is not understood. For example, the product supplier wants to understand your requirements so they can design the technology layer. You want to understand the global, regional and local influences to inform the design. Not understanding your current state can delay the answers to the questions the product supplier has asked while you obtain the answers, and this can result in tension between you and your supplier.

Preparation for organisational change and adoption

The more you know about your organisation, the greater the chance the changes will be adopted. Success will be driven by your organisation's commitment to understand, own and drive change. When you get this right, WFM initiatives will be embraced and adopted.

Education and methodology

WFM and its functional components, such as rostering, are not a formally recognised discipline or skill. Most people who work in WFM end up working in the industry due to circumstances. There are a number of technology courses (from suppliers) that define how to access, design and configure a WFM solution, but don't explain what this means to your organisation.

More recently, we have seen the emergence of private institutes and associations looking to bridge a gap, but there is not yet anything in the market that fills the gap. For me, the real point is that the WFM discipline is yet to be adopted by universities and certified education institutions.

Why do I call this out? It really comes back to the key reasons I wrote this book - to increase WFM awareness, look at WFM through a business lens and provide a repeatable, practical methodology. Once you have this, you increase your ability to give the greatest value back to your organisation and the people who work in the area of WFM.

As part of my commitment to WFM globally, I will soon launch the WFM Association to help provide knowledge and education back to the industry. More about this at the end of this book.

Top take outs

- → Understand the rostering model(s) that are required.
- → Will rostering be centralised, decentralised or a hybrid?
- → It's never too early to start cleansing data. The cleaner the data, the better informed your decisions will be.
- → Many factors influence the selection of a supplier, especially industry fit, cultural fit and ongoing product investment.
- → Supplier accelerators can push the responsibility of the implementation and achievement of business outcomes back to the organisation.
- → Never lose sight of the outcomes you are looking to achieve.
- → Award interpretation is a high-risk area; don't avoid it, simplify it or push it to the side.
- → The internal investment in people to ensure success of a WFM initiative is key to delivering a successful outcome.
- → Understand what flexibility you can allow across countries, regions and locations.
- → Create organisational awareness and build education from the beginning.
- → Process change will be required to deliver the outcomes.

Where to next?

This chapter covering Step 2 - Prepare, focused on the determination of business and supplier(s) readiness to enable a successful implementation. We now move on to Step 3 - Implement, which focuses on the implementation of technology and the business change that is about to occur in your organisation.

1 M. Markovski, 2018, Apps Run the World, Top 10 Workforce Management Software Vendors, Market Forecast 2016-2021, and Customer Wins', 8 January 2018. viewed 17 January 2018, < https://www.appsruntheworld.com/top-10-workforce-management-vendors-market-forecast-2016-2021-and-customer-wins//>.

2 Fair Work Commission, Awards & Agreements, viewed [2 August 2017], <https://www.fwc.gov.au/awards-and-agreements/agreements>.

Chapter 9

STEP 3 – IMPLEMENT

Determining the business outcome you are looking to achieve is in many respects the easy part. Working out what you need to do in your organisation to achieve the outcome is often the harder challenge. Far too often I have seen a sales process finalised with the implementation literally starting the next day. This quick start is fine, provided you are prepared for the task at hand. In my experience, the preceding step, Step 2 - Prepare, is rarely completed. This compromises the balance between fully understanding your detailed business requirements to achieve the business outcome and meeting the technology supplier work effort.

What you need to know to adopt WFM

Step 3 - Implement - focuses on making sure the people, process and technology are optimally planned so that everyone in the organisation knows what to do and when to do it. This includes understanding how the organisation is currently operating, creating standards and rules for the implementation, understanding the scope and plan, determining what methodologies will be used, how the delivery will take place and how change/training/support will be enacted. The remainder of this chapter unpacks the key considerations to ensure the implementation is a success.

Figure 9.1: Smart WFM Methodology: Step 3 – Implement

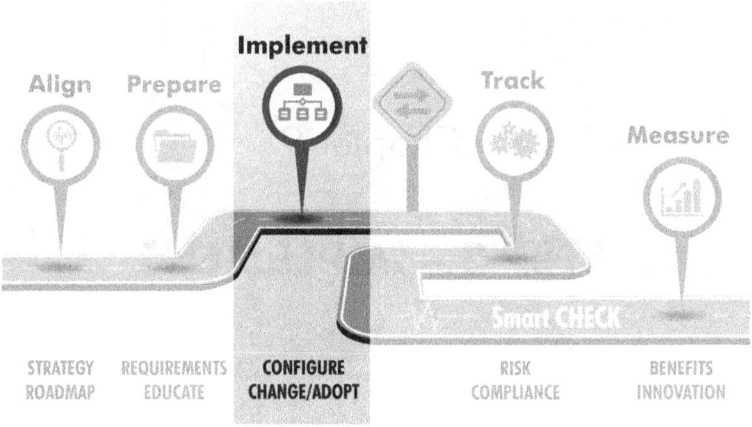

Survey

The more information you have about your organisation, the better prepared you will be for the changes ahead. WFM projects have a significant impact on the operations area of your organisation. Over time, it is common for operations management and staff to develop their own ways of working. This is not a bad thing, but this diversity can influence the requirements, design, change and adoption associated with the implementation of your new solution.

An ideal way to find out more about your organisation is to complete a structured survey prior to commencing the implementation. The more knowledge you have, the more informed you will be in the upcoming decision-making process and this will accelerate time-to-value.

Whether you run a large, multi-faceted business or a smaller, localised business, the process to complete the survey is similar, but the time required to complete it may be different. The survey process can benefit from a central point of reference to drive consistency in the process to collect the data. The more accurate your data, the more accurate the decision-making process.

Key areas of required business insight to consider for the survey are discussed below.

Sites and locations

For each of your sites, understand who are your key contacts in operations, HR, payroll, finance and IT. In larger, multi-faceted organisations, variances are common, so be sure you know who can support you to obtain the required information. It may seem elementary, but knowing who to contact at a site to help you work through implementation items is invaluable.

The type of questions you need to consider are:

→ How many sites are there? This will help you to work out how many operational methods there are and, therefore, the size of the changes required.

→ Are the sites spread across multiple geographies? This will help inform you of things such as the reach your project team needs to determine the requirements and to support the solution.

→ Does your organisation own the site(s) or are you subcontracted to work there? This will help determine the physical and technology access protocols you may be required to abide by.

→ What technology does each site run; what browser; what is the network speed? This will have an impact on the desktop/mobile technology versions that you can use to run the software.

→ What HR/WFM applications does each location currently run? This will help decide things like how many integrations will be required and the quality of data.

The answers to these questions will provide relevant input in upcoming decision-making.

Employment type

It's important to understand the basis of employment for your people at the sites and locations. Are the staff:

- → full time
- → part time
- → casual
- → contractors/gig workers
- → seasonal
- → salaried?

How many people fall under each of these different employment types? These breakdowns will help inform operating decisions later in the implementation. For example, will the onboarding process be the same across employment types? Will the payment conditions be the same across employment type? Do you need to communicate change differently to each group of employees?

Current method of rostering

Understanding how the site/location sets rosters will help to inform the future state. For example, a hospital may have two sites in a major city. Within each site there may be multiple locations: the emergency department, radiology, operating theatre and so on. The number of staff and the hospital layout may impact the approach to rostering. This will help to inform the future-state rostering models we identified in the section on types of rostering in Chapter 8. Knowing the current state will help to determine the future state and, in particular, the changes required.

When taking a closer look at your operational models, you often find subject matter experts who can become an integral part of the project team. These experts are generally very passionate about what they do, and they can have an extremely positive influence on future operating models and how well people adopt the changes.

Anecdotally, I often find that when you staff your WFM project with business subject matter experts, the credibility and acceptance within your organisation is greater. It's always better to hear from someone who has learned from real experiences within your organisation than from an external consultant.

Award catalogue

Understanding your awards, the history of them and the way they are interpreted can be one of the greatest challenges/risks associated with any WFM implementation. I spoke in the last chapter about award interpretation and how awards are not one size fits all. This section should be read in conjunction in the previous chapter.

This step in the survey process is a key step to mitigating many of the award risks that have been previously identified. You would be surprised how few organisations are able to provide a full list of agreements they have in place, across which sites/locations they are applicable, who is responsible for interpretation (one or many people) and what is the review date for that award.

When you engage your product configuration specialists, they will ask questions to enable configuration of their products. If you are not able to answer these questions promptly, it can place significant stress on people, timelines and scope.

Time-collection methods

Understanding your current process of time collection is important for a number of reasons, the two main ones being:

- → you can determine your baseline: paper, electronic, or a combination of both methods
- → it informs you of the current way of operating and enables you to understand process change.

With larger, multi-faceted organisations, it is not uncommon for multiple time-collection methods to exist. You may also find there are bespoke time-collection methods that solve business problems you may not have been aware of. If you remove or change the current time-collection process, the business may lose mission-critical functions.

Sample artefacts

Obtaining sample artefacts is a great way to capture a current snapshot of your business operating model. I've always been a believer in 'look and learn'. One big advantage of collecting these artefacts is that it enables you to see what is being collected. You often find things outside your planned scope that are integral to the business operating smoothly. As these systems are operational in nature, carefully consider the implications of any changes on the end-user.

> **For example:** Thompson Road Repairs finds out that its gangs are using five different materials when repairing the road. Information about which material and how much of it was used in the repair was being collected on the same form and at the same time as employees were completing their timesheets. The data was used to facilitate materials management and procurement activities. The project team learns that other parts of Thompson Road Repairs require this, and without this information the business will not operate effectively. This knowledge does not mean the project has to cater for this data collection via technology; rather, it needs to cater for this scenario as part of its go-forward operating model.

The earlier in the process you identify scenarios such as this, the quicker you can find an optimal resolution or workaround.

Data cleanliness

Data analysis will help determine which sites and locations are more thorough and diligent than others. For example, are you capturing accurate actual data once; do you have to manipulate the data to represent actuals? The answers to these questions will help drive subsequent reporting requirements and business KPIs.

With the emergence of AI and predictive algorithms, the cleaner the data the more accurate predictions will be. If your WFM system is predicting a late arrivals to work with 95 per cent accuracy and the data to predict this is out by 15 minutes every day, the prediction will also be out.

Knowing which sites, locations and people drive clean data will allow you to propagate these good habits throughout your organisation.

Survey key learning

Those organisations that complete the survey process thoroughly lay a solid foundation for their future operating model. Conversely, those organisations that assume they know everything already are at greater risk of building on a shaky foundation. It is challenging work to complete this type of thorough analysis, but the results speak for themselves. You will find yourself referring to this key information over and over during the process.

Detailed requirements

The previous two steps in the methodology have a strong focus on the alignment of your organisational goals to your workforce goals. Now you also know from the previous section the importance of understanding how you currently operate.

At this point, you can take your big picture and overlay it with the current way of operating to create your new future work model.

I believe this is a crucial step in any project as it allows the 'rubber to hit the road'. Everything you do from this point forward will be fundamentally impacted by these decisions. While you are still in the implementation stage, you can make changes to the design using information at hand, but once you go past this step, any changes to requirements may have a fundamental impact on the benefits your organisation will receive and when it receives them and the overall costs to deliver. These requirements will influence your operating model, which benefits you can deliver and how quickly you can deliver them.

Requirements baseline

Your current level of organisational knowledge will allow you to create a requirements baseline. If you have different rostering models, award conditions and digital requirements, these will all inform the requirements.

Many service providers and suppliers will have a requirements' baseline they can bring to the table so you don't start with a zero base. If you have organisational goals and workforce goals to refer to, you will be better able to prioritise the requirements.

Process baseline

This is a logical follow-on from requirements. Once you know your requirements, you can define a process to meet those requirements. Creation of a process baseline has several benefits:

- → You are able to consider the end-state solution from a user perspective, i.e. how the user will complete their job on a day-to-day basis.
- → The requirements will be tied together to form a process, i.e. you won't look at the requirements in a singular, non-connected way; you'll look at them from an end-to-end perspective. Often this way of looking at things is referred to as a persona.
- → If you have a defined process, you know that once followed, it will help you to achieve your business outcomes.
- → Similar to the requirements baseline, many of the service providers and suppliers will have a process baseline that they can bring to the table so you don't start with a zero base.

Traceability

Traceability is a way of ensuring that you are meeting your business outcomes and mitigating the risk of the technical solution failing to meet the business objective. Traceability flows in a top-down manner from:

- → organisational goals;
- → workforce goals;
- → WFM benefits;

- → WFM enablers;
- → requirements;
- → processes; and later to
- → test cases.

There are also non-technical considerations outside traceability you need to consider such as employee and customer experience and other benefits we defined earlier in the book.

Business impacts

From a business perspective, don't rely on third parties completely to own the implementation process. There are a few reasons for this:

- → You ultimately need to take responsibility for your own destiny.
- → Your people will only truly accept and respect a solution that is driven by you, i.e. your people don't want to be told by product suppliers or consultants what to do.
- → You are going to change the way you operate so you need to understand what this means.

More and more I see organisations successfully using an external supplier to drive and deliver the implementation outcomes and they work together and govern the overall direction. This process works well as the external supplier will use their WFM experience while you will provide your deep business knowledge. The inputs and skills are highly collaborative.

Process

The amount of process change depends on your starting baseline. At one extreme, you could be starting with paper time collection and rostering processes and you are automating them. At the other extreme, you could have an on-premise WFM solution with efficient processes and are migrating to a cloud solution.

Most organisations will be undertaking process change. You will only achieve the business benefits if you follow the process.

Supplier accelerators – Part 1

Many suppliers will provide templates and accelerators to speed up your implementation. While this reduces the effort required by the supplier, the effort required by your business team may increase.

The suppliers provide the templates giving a brief explanation of how to complete the template. What is generally lacking in this education are all the associated implications on your people, process and technology. For example, a supplier template might ask you to define the locations that you roster from. What you don't realise are the technology implications of this, such as:

→ the data control areas;

→ how functional security is applied by location;

→ reporting roll-ups occur by location; and

→ locations are required before the system configuration can take place.

Similarly, if you ask the supplier for help outside the initial education session, you need to book the time of the consultant to answer your questions. The time to do this will be charged on a commercial basis, and the implementation can suffer due to the associated turnaround time.

I am not saying accelerators are a bad thing but, ultimately, they transfer the responsibility of completion to you.

Technology

We're back now to a discussion on technology. This time we're interested in where it fits in the implementation step.

Configuration standards

Prior to commencing the configuration of any solution, you'll need to understand from your product supplier what standards they are going to adopt to build the solution.

The larger and more multi-faceted your organisation, the more important this becomes. If you share too much configuration, a change to one site's requirements in the future may result in changes of configuration to another site, even though their requirements may not have changed. This can result in a regression test, which can be costly in terms of integrity, time and money.

Consider configuration standards in a top-down manner:

- **Global** - what standards will be adopted that apply across the entire solution, e.g. naming conventions, costing rules and reporting roll-up?
- **Regional/business unit** - what standards will be adopted that apply at a regional level, for example EBA set-up, the rostering process, time collection process?
- **Local** - what standards will be adopted that apply locally, i.e. rounding rules, site-based allowances and deductions and the rostering set-up?

Not making these considerations prior to commencing configuration can have a significant impact on your WFM solution's ability to flex and meet the demands of changing business requirements.

Enterprise architecture

Understanding where the WFM solution fits into your organisation's overall enterprise architecture will help determine the boundaries and crossover points with other applications. From a WFM perspective, there are several components that make up an enterprise architecture. Here are the main ones.

Applications architecture

What applications exist in the organisational landscape? Is your organisation part of a group entity where there are standards in place around applications that should be followed?

WFM has strong integration with HR, payroll, finance and, in some industries, job management and forecasting systems. There are many permutations of which system functions are completed in what systems to deliver an optimal solution. Figure 9.2 gives a simplified sample architecture from a WFM systems perspective that could be tailored to your organisation's unique situation.

Figure 9.2: Sample applications architecture

Systems and functions

HR

HR org structure
Onboarding
Off-boarding
People master
Skills and certifications
Traditional self service

Payroll

Payment ($)
On-costing
Pay-based reporting

WFM

Time collection
Rostering
Forecasting
Availability
Shift management
Exception management
People costing
Hours interpretation
Reporting
Access control

Finance

General ledger
Financial forecasting

Job management

Job creation
Job dispatch
Job costing
Job reporting

Forecasting systems (industry dependent)

Points of sales
Patient admission

The four main design questions I see debated are:

1. What system will leave be completed in?
2. What user interface will be used for mobility?
3. Where will self-service be completed?
4. Where will award interpretation be completed?

My recommendation is to deal with all these questions at the outset of any initiative. When determining the answers, ensure you consider what this means to an end-user and whether the system has been built by the product supplier to deliver those specific end-user requirements. You need to understand what functions reside in what system. WFM systems by their nature are designed for the end-user first.

Many organisations also have a phased/transitional approach to deliver the end state; they move from current state to interim state and final state. This helps them to keep the initiative moving while balancing the long-term needs of the different vested parties.

There is increasingly functional crossover between HR and WFM systems. I think this will continue and, ultimately, we will end up in an industry consolidation phase where these technologies will be offered by a single technology supplier or tightly coupled technology suppliers.

We are also seeing more organisations using best-of-breed point solutions for specific functions. There is more about this in the upcoming section on platform architecture.

Technical architecture

Understand your technical requirements early in the implementation process. Engage with the IT team to identify any standards they have in place and work with them to design an optimal operating technical architecture.

There are many considerations across areas such as security (data and functional), network, servers, application, integration, desktop and performance - and items across design, procurement, commissioning, testing and ongoing support.

From a project implementation perspective, there is a changing focus on technology with cloud. While many of the on-premise considerations still stand true, the focus on individual areas may change. For example, with cloud, many of the server considerations are removed.

Platform architecture

Some product suppliers are building their applications on an open application processing interface (API). This means that it is possible for you to send and receive data to/from almost any complementary platform. In turn, this is seeing best-of-breed platforms being selected to achieve specific functions, e.g. onboarding, people master, rostering, payments, pay and so on.

WFM platforms then require a number of other technology-related questions to be answered around application security, integration, consolidated reporting, user interface, etc.

If you are going to adopt an open API and platform architecture, consider working through the technology-related questions and ratify this architecture early in your WFM initiative.

For further information refer back to Chapter 5.

Mobile architecture

There is an increasing trend to design mobile-first systems. In Aron Ain's interview he said the strategy *is* mobile. These days, any business function you undertake, you generally do on a mobile device. Some suppliers excel at this and others struggle. Many organisations are trending to bring your own device (BYOD) solutions. This means there are myriad mobile operating systems that the technology needs to be capable of working on. Not every product supplier has their application suite working seamlessly across all device technologies. Here are three key things you need to consider from a user experience perspective:

1. If some devices are not capable of running the WFM application, how do you achieve a single process across your organisation?
2. Do you now mandate that some functions need to be completed on another technology such as a PC?
3. Are multilingual capabilities available across the mobile devices/ operating systems?

Understand whether you are in this situation early in your implementation, as working through the resolution can take some time.

Process architecture

Earlier in this chapter, I spoke about organisational benefits being delivered by following a defined process. Process also needs to be carefully considered from a technology perspective. For example, if you complete a business process across a number of systems, will the outcome be usable from an end-user perspective? Designing technology process architecture will help to mitigate the risk around this.

Data architecture

The greater the number of system and technologies in play, the more important data architecture becomes. Ultimately, the data that resides across your WFM and broader HCM solutions will be required for operational, management and predictive analysis. Understanding where data resides, where is its source, what systems require it and at what frequency, becomes critical.

Time-collection architecture

An area to consider early in WFM implementations is the installation of time-collection devices. Technology today provides far greater flexibility for this, via wired ethernet, Power over Ethernet (PoE), satellite and mobile networks.

Some organisations have staff working on client sites under contract. For example, a cleaning company may supply staff on a permanently contracted basis to a hospital. In this case, the hospital may not allow the security company access to their network to collect time information. While there are several options for the security company to solve this challenge, working through the permutations of options and coming up with the most efficient, cost-effective answer may take some time.

Deciding where to install the device may also need some detailed consultation. In heavily unionised sites you may need to go back and forth between the parties to determine a suitable location. Take a situation where the placement of a time-collection device has been deemed to negatively impact pay, for example if the time-collection device is moved from the front gate to the production line, employees' clock-on time will start later, potentially affecting their pay.

Supplier accelerators – Part 2

As noted earlier in this chapter, suppliers will often provide accelerators and templates to speed up your implementation. Similarly, once these templates are completed, the information contained in them will be uploaded to a 'pre-configured' template and the solution will be technically ready to switch on (provided your data is correct).

The suppliers will provide education on how to use the solution. What is generally lacking in this education are all the associated implications for your people, process and technology. For example, the new process might mean that staff can set their availability. This can result in scenarios such as:

- → Department A wants employee-based availability, but department B does not. The template allows all departments to set their availability.
- → Availability preferences can only be set on weekdays and not weekends. The template allows availability to be set for all days of the week.

If you ask the supplier to better understand these processes in relation to your business, you need to book the time of the consultant to help answer your questions. The time it takes to answer these questions will be charged on a commercial basis, and the implementation of the changes will increase the project timeline and budget.

Ultimately, you need to modify your business process to fit the technical solution, or understand that personalisation to your business requirements will have associated impacts on the timeline and your costs.

The people involved in the project

It is people who drive the success of your organisation's WFM implementation. In this section, we place an emphasis on the people considerations from a project perspective.

Project team

WFM works across a number of departments within your organisation. There is a high chance you will be implementing process change to achieve the desired business outcomes.

You need to select a project team whose members have an ability to think objectively and consider the overall greater good of their organisation, their people and their customers. People are at the core of these initiatives, so people-focused team members are key.

Project roles

I previously introduced people considerations for a WFM initiative in Chapter 2. This section should be read with those considerations in mind.

For many project team members, it is often the first time they have worked on a project of this type. Once you commence the implementation step, project team members require clear direction on the role they play in the project. Ensure the team members know your expectations of them and their responsibilities.

Organisations often put KPIs in place to help drive the right outcomes and behaviour of project team members. In some circumstances, I have seen the role of the project team member become redundant at the conclusion of the project. In these cases, having a KPI attached to an incentive, often financial, is critical to ensure a successful outcome.

Project team education

It is a good idea to have some form of induction planned for your project team. Help the team members understand what the objectives are for the project, their roles, how to handle issues, risks, time-recording requirements, work hours, site access, remote access, weekly reporting and so on. A lot of time can be wasted if these items are not covered before the project gets underway.

External consulting teams

Most WFM projects have external consulting teams that contribute to the project's overall success. The closer and more corroboratively you work with these teams the better. It's important to realise that everyone on the project is there because they have something of value to add. Take time to understand the background and specific skills of each member of your project team. It's rare you don't make new friends and gain new knowledge in these circumstances.

I have seen situations where the project teams end up working in an 'us' and 'them' scenario. If you find yourself in a position where this is happening, ensure all the parties address this openly in a constructive way. Project governance input may be required at this point.

Remote teams

Remote work is becoming common practice. As video conferencing has matured, and cloud-based collaboration tools are generally more promptly available to remote team members, this way of working will continue to grow.

There are often commercial benefits to be gained from working this way, as it minimises travel and other expense costs.

Scope and plan

Prior to commencing the implementation and bringing on board the delivery team, make sure the scope is clearly articulated and understood by everyone. It doesn't matter if you are using agile or waterfall delivery techniques, be clear on what you are delivering and the plan to deliver it.

Methodology

The Smart WFM Methodology is in no way meant to replace traditional or supplier methodologies. This methodology is designed to complement existing methodologies by providing experience, knowledge, anecdotes and examples of what I've seen occur in organisations. You can plug and play any of these learnings into any methodologies.

The Smart WFM Methodology is designed to bridge the gap between existing methodologies and the additional nuances that come into play with WFM initiatives. There may be aspects I have not documented that are relevant to your WFM initiative. Feel free to adapt the methodology to your organisation's particular circumstances.

Traditional methodologies

I refer to traditional project management methodologies as being the likes of PMBOK and PRINCE2. These methodologies were created after years of learning across the delivery of many projects.

The key thing to realise is that these methodologies are generic in their nature. They don't take into consideration supplier methodologies or the nuances of rolling out WFM. A notable example of this gap is the parallel testing process, which we will look at later in this chapter. I do not suggest you discount traditional methodologies; in fact, the governance and thinking they provide is very rigorous.

Supplier methodologies

Many product suppliers have their own implementation methodologies. These methodologies have generally been created after years of learning across their implementation process.

The important thing to note is that these methods generally place a stronger emphasis on the product implementation process, not the broader implications on your business outside the product, such as business change and adoption. You may need to demonstrate you have followed these methods to obtain ongoing vendor technical support.

Agile and design-thinking

Adopting agile and design-thinking delivery techniques will allow you to consider the end-user and get runs on the board quickly.

These concepts are gaining momentum and I think there is a lot we can learn from these delivery methods. When was the last time you used a really cool app? What did you like about it? I anticipate you would say it was easy to use and solved some burning problem. Chances are this app was developed using agile and design-thinking methods. I think WFM and HCM in general can learn a lot from the software development community; they have been delivering efficiently using these techniques for many years now.

The beauty of agile and design-thinking approaches is that you continue to learn from your real-life, business-centric experiences.

Waterfall

Consideration of agile versus waterfall delivery methodology should also be considered. Refer to the previous section in Chapter 6, Get runs on the board quickly - Agile and Waterfall.

Data

WFM systems rarely operate in isolation. At a minimum there are two other systems - the people master system feeding WFM and WFM feeding the payroll system. It is more common for there to be other systems also sending/receiving data to/from WFM. For example, retail has POS, forecast sales; service organisations have job management; and the health sector has patient intake demands.

Data alignment across these systems for any end-to-end testing can take considerable time and effort to coordinate. Understand the data requirements early and get a good handle on what is required to align data. Understand the effort it takes to obtain/create data and to refresh data at a point in time- for example, understanding the backup and restore criteria. Take copies of data at appropriate points in time so you can restore it as required to complete your testing process.

As noted in the data cleanliness section of this chapter, with the introduction of AI and machine learning, clean data is essential. You can only measure and learn from what you collect. You can only learn correctly if you learn based on correct data. This is no different to a student being taught the wrong thing by a school teacher.

Historical data migration

There are two main types of WFM data: historical time data and rostering-related data.

I have rarely seen historical time data migrated when an organisation is moving to a new WFM platform. Often discretionary decisions are made that implicate the payment results. Not knowing what led to these discretionary decisions and why they were made can result in award-interpreted hours being different between the old and new systems (manual or electronic), resulting in reconciliation anomalies.

It is more feasible for rostering data to be migrated to a new system in terms of the baseline rosters that are required. To do this, import and export functions at the technology layer are required. Ask yourself: will rosters in the future system be the same as the current system? If there is a difference, migration may not be possible.

Testing

Testing in any project is a critical activity. In particular, the testing of award-interpreted hours and its integration with payroll to produce the correct results is of the upmost importance. In fact, if this step fails, the result can be almost catastrophic, with people not getting paid correctly, and it can directly impact people's livelihoods.

What can often be found at this point of a WFM project is that what has been configured works according to the documented EBA, but it does not reconcile to the actual payslip. Reconciliation can be time-consuming and can impact negatively on the delivery timeline. This scenario is common and the risk of it occurring should be addressed in the project planning process.

Parallel runs are common in the deployment of WFM and payroll solutions. Parallel runs are generally not seen in traditional methodologies but are sometimes seen in product supplier methodologies. There are two types of parallel: point-in-time parallel and actual parallel. Having previous experience when planning and executing either type of parallel run is invaluable in this area.

Point-in-time parallel

These parallels are very common on payroll projects. This involves running payroll for a pre-determined previous pay period and simulating that pay period's results in the new solution. While the description is straightforward, the permutations and nuances associated with planning and coordinating are not.

If you are rolling out a new WFM and payroll solution at the same time, the challenges are further increased. There are often discretionary decisions made in the WFM system that influence pay results. If you don't know what led to this discretionary decision, there may be a reconciliation anomaly. Reconciliation anomalies usually have two main causes:

1. The site is paying differently to the EBA
2. Discretionary decisions are being made, such as manual rounding, payment of overtime, variability in the manager application of rules.

Actual parallel

Actual parallels occur when you run the existing process (manual or automated) at the same time as the new automated process, i.e. you have two systems of record for a defined period. This type of parallel can be good when you want to bed in a new process and educate your team members in a pseudo-live real environment and compare pay results in real time.

Many factors influence this type of testing, including the number of users, ability to support the users and ability to deal with configuration changes in a real-time environment.

Change

Change is quite possibly the single largest impact associated with any WFM project. As I have noted previously, WFM projects have people at their core. Delivery of business benefits requires process change and, subsequently, people change to achieve the outcome.

Below are a number of roles I've seen create effective change in organisations. Don't get too hung up on the title of the role; rather, focus on the key tasks that the role is enacting.

Stakeholders

When senior management takes proactive involvement, there is a better chance of achieving rapid success. It's important to make sure your stakeholders are at the right level. For example, if you are part of a multi-faceted organisation, and you are deploying the WFM solution over multiple business areas, the key stakeholder needs to be at a level in the organisation where they have reach, influence and leverage over the different businesses. In this example, constant involvement and support from the CEO may be required.

Champions

The champions are your people who buy into the organisational goals, workforce goals and understand and accept the changes that will take place. The champions are generally operational managers within your organisation who have the strong respect of their teams. Champions must be able to canvas support and mitigate resistance.

Key users

The key users are your people who see the big picture and have a hands-on role. They normally embrace the solution and understand the benefits it will bring the organisation. They become the go-to point for the day-to-day operational issues.

Your people

Your people are those at the coal face day-to-day; they adopt the new way of work as part of their daily routine. They use their mobile phone to see when a shift has been allocated, use a time-collection device to record hours worked and so on.

Communications

The earlier in the process you can communicate why there is change, the less chance of guesswork in your organisation. I have found that the delivery of these messages is most effective from those who are part of your organisation and who are respected by your people.

Commencing a project with a meeting that involves everyone from stakeholders through to a representation of those at the coal face is beneficial. In this meeting, clearly indicate the business benefits and the key changes that will be coming. Have a transparent mechanism in place to solicit and communicate feedback regarding key items of concern.

Business role changes

In Step 1 – Align, I introduced people considerations for a WFM initiative. This section should be read with those considerations in mind.

With the introduction of WFM, almost certainly you will see adjusted or new roles within your organisation. Align job descriptions to this new way of working early. You may need to consider the introduction of new KPIs in order to align the organisational goals to workforce goals to ways of working. This may be necessary over a number of roles.

It may also be necessary to roll out and negotiate pay changes caused by EBA adjustments associated with the implementation.

A final note: even if everyone believes in the solution, if it is not usable it will not be adopted. Be careful not to lose sight of this in your implementation.

End-user training

A common question on WFM projects is how and when should you train your end-users? While the answer to this will depend on your organisation's circumstances, there are a few considerations that may help to answer this question.

In-product

Many of the product suppliers are now moving to 'in-product' training, i.e. the training materials and processes are built into the product. This has many advantages from a time, cost and acceptance perspective. Training is available from the moment the product is turned on and can be updated and referenced in real time. The suppliers provide this training and it remains current as new product releases are made available. This is particularly common in cloud environments.

As each day goes by, we are getting closer to training that is essentially self-learned, i.e. the system is logical to a point that the end-user can teach themselves. Take, for example, Facebook and Instagram, where new functions are added and users adopt the changes naturally.

Short videos

These are generally short recordings lasting one or two minutes that demonstrate how to complete a specific task. Often they are taken in real time with an end-user explaining how they use the technology.

Web recordings

Similar to short videos, but instead using a screen-flow capture tool, web recordings are used to demonstrate the system usage. Narratives can be added over the top of the recording to further enhance understanding.

Cheat sheets/quick reference guides

This method is still popular and tailored specifically to the task at hand. With such a strong focus on mobile, the delivery method of all training should be mobile friendly.

Go-live and support

Consider how much staffing you will require to support your WFM solution through your go-live. The change roles that were defined above can be retained throughout your support period. The answer to how much staffing you will require depends on the amount of business process and role changes that have been made.

You will require a support structure for each site. Many organisations also skill up their internal helpdesk (either with internal staff or consultants) to help everyone through this support period. From experience, this period generally lasts for between two to four pay-runs per site, assuming your solution is stable.

Having daily meetings, sometimes known as stand-ups (on-site and/or remote) with your stakeholders, champions and key users is valuable. Everybody gets to hear promptly what is working well and what is working not so well. Learning as a group can be powerful.

Make sure you are truly ready for the change. Consider the fundamental wisdom to not bite off more than you can chew and considerations around all business areas going live at the same time using a big-bang go-live approach. I challenge you to think through the implications of this and make sure you are ready for it. Consider the implications from a business operating model perspective and from a people payment (payroll) perspective. While both these areas are complementary, they are quite different. The operating model impacts process and ways of working. Payments implicate what goes home in the pay packet each week. Both models take different skills and staffing to support.

Ongoing support

Once the project has concluded, who will support your WFM solutions in the future? You have the option to staff this internally, use external consultants, or a combination thereof.

The complexity in configuration with WFM solutions is often around the assessment and configuration of the business rules, in particular the award rules along with integration components. Many large organisations pay careful consideration to upskilling or recruiting people with these skills. If the skills are used regularly and there is good variability in the role, it's a viable option. People with these skills will often lean towards working for the product supplier or consultants due to the variety of work they get to perform.

Top take outs

- → Understanding the current way of working will allow requirements and change to be understood, enabling adoption of the new systems.
- → Sample artefacts are a great way to understand what is in place and how great the change will be.
- → Taking ownership of your implementation process to achieve your outcomes is key.
- → Supplier accelerators push responsibility on the organisation and don't always enable a full understanding of the business implications of their usage. They are often configuration documents from a supplier perspective (not a business requirements document).
- → Create an enterprise architecture to baseline your solution and to create delivery guidelines.
- → Select a strong project team and set clear expectations about roles and responsibilities, and set KPIs.
- → Clearly understand the business outcomes, scope and plan. Relay this to the project team regularly.
- → Determine how the supplier methodology and traditional methodologies will be utilised.
- → Determine how agile and/or waterfall delivery methods will be utilised.
- → Create a flowing change, training and support network.

Where to next?

We've now covered Step 3 – Implement, focusing on determining the steps necessary to complete a successful implementation. The next chapter introduces Step 4 – Track, which looks at the tracking processes to keep your implementation pointed in the right direction.

Chapter 10

STEP 4 – TRACK

I have seen numerous governance methods applied across WFM initiatives. There are specific nuances with WFM initiatives, and if you have not experienced these before they can stymie the goal of achieving success. The approach to testing for correct pay results and determination of the best approach to go-live are often hot topics and play an integral part in determining the success or failure of your initiative. Similarly, how to interpret an EBA seems straightforward, but once you scratch beneath the surface you can often realise some unexpected business interpretations. WFM implementations result in high-impact business change, which can lead to highly emotional outcomes with those involved. Having the right governance balance in place to drive business outcomes is of the utmost importance. This fourth step in the Smart WFM Methodology will help you manage risk and ensure compliance (see Figure 10.1).

How to keep your WFM initiative moving successfully

Step 4 – Track focuses on identification of the typical items I consider important to ensure success of your WFM initiative. The points I raise can be used as a checklist of WFM-related considerations. The topics discussed in this chapter and the other chapters in this book should be considered to complement your selected governance method. Also refer to Chapter 6, The fundamentals to ensure success in any WFM initiative.

Figure 10.1: Smart WFM Methodology: Step 4 – Track

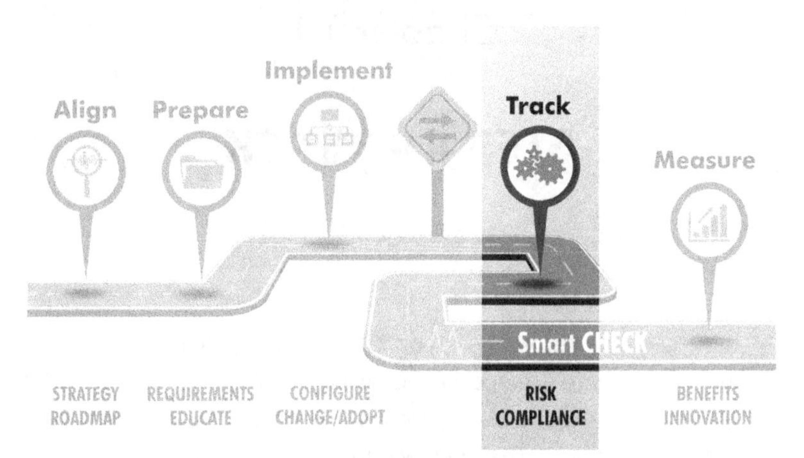

Benefits and business case

The benefits you are expecting a WFM implementation to bring to your organisation, which you outlined in your business case, should be supported by your project governance activities. They will be constantly reviewed and challenged as you move through the delivery program.

You will have locked down the benefits and business case at the outset of the WFM initiative and agreed relevant items with all key stakeholders on the project, the implementation team and those people at the operational coal face. If you make the benefits and business case visible, it will enable decisive decision-making throughout the project.

Understanding the benefits and business case can make or break your WFM initiative and give it a solid foundation on which to build. It allows a link to be established between the benefits to the organisational and workforce goals.

Requirements

Requirements flow from the benefits and are aligned to your organisational and workforce goals. The requirements will be created at a high level in Step 2 - Prepare: What you need to know before you start a WFM initiative and finalised in Step 3 - Implement: What you need to know to adopt WFM.

By making the requirements visible with key stakeholders on the project, the implementation team and those people at the operational coal face, you will provide clarity about what you are delivering and remove unnecessary noise.

Current state

Understanding your current state is an important step to inform your detailed design, change and adoption activities.

Many organisations, especially larger, multi-faceted organisations, sometimes do not undertake this step and plan to move straight to a new way of operating. While this approach can work, if you follow it, you run the risk of not fully understanding your operating model. You may miss requirements and fail to understand the change and adoption activities required to make the project successful.

Technology

How does technology help in the tracking stage of your WFM initiative? Let's take a look.

Configuration standards

It is important to have configuration standards in place prior to commencing configuration. These standards enable you to put in place the correct building blocks to allow the configuration to be completed and maintained consistently across your implementation and support team, mitigating the risk of rework. Refer to the earlier section on configuration standards in the previous chapter for additional information.

You need to agree on configuration standards with your WFM supplier and ensure the implementation team understands them. Configuration should not commence without these standards being in place.

Localisation

In Step 2 – Prepare we covered global, regional and local influences that could affect your WFM initiative. Once you decide what will be addressed at a global, regional and local level, you need to keep track of it. Consultants and business representatives are notorious for deviating from these rules. Pressure will often come at a local or regional level to implement a specific request outside the agreed governance process, for example, you may be approached to tackle a legacy EBA rule or roster a new group of employees. If you don't keep a close eye on this, you could quickly end up with a solution that does not comply to your standards, will become more costly to support and will not deliver the agreed outcomes.

Enterprise architecture

Your enterprise architecture is technical in nature but it's important to ensure your systems work in a unified way. If these items are not covered, you may find your solution is difficult to use, not intuitive, or has multiple or manual entry sources for information.

Remember there is a number of components that make up your enterprise architecture, namely application, technical, platform, mobile, process, data and device architecture.

Agree on the enterprise architecture with key stakeholders of the project and the implementation team. When the enterprise architecture has been thoroughly considered and fully understood, you will be confident that your technical solution will deliver your business benefits.

Testing

We covered historical data migration in the previous chapter, but to recap, there are two main types of WFM data: historical time data and rostering-related data.

I have rarely seen historical time data migrated when an organisation moves to a new WFM platform. Often discretionary decisions are made that implicate the payment results. Not knowing what led to these discretionary decisions and why they were made can result in award-interpreted hours being different between the old and new systems (manual or electronic), resulting in reconciliation anomalies.

It is more feasible for rostering data to be migrated to a new system in terms of the baseline rosters that are required, for example shift start and stop times, demand forecasts and jobs to be completed. To do this, import and export functions at the technology layer are required. Ask yourself: will rosters in the future system be the same as in the current system? If there is a difference, migration may not be possible.

Be clear on what you are going to test and how you are going to test it. There are two major considerations from a testing perspective: (a) pay results and (b) your operating model. Consider the requirements for each type of testing carefully to ensure you mitigate the risk of making incorrect pay calculations and adversely affecting your operating model.

The project board and governance

The project board members play a key role on your WFM initiative as they keep it pointed in the right direction. I can't stress enough the importance of active participation in matters of governance by the project board members. Some of my key learnings in this regard follow.

Representation

As we have discussed already, WFM initiatives impact people. WFM projects cut across multiple areas of your business, e.g. operations, finance, payroll and IR.

I have seen many occasions where the suppliers are not represented on the project board. This always rings an alarm bell. I am not saying your product suppliers need to be across all your financials and internal/confidential business decisions, but there are some things they do need to be across.

Take your project's timeline, for example. The time it takes to configure the selected technology is best provided by the supplier, so it's no use setting a configuration timeline without input from the supplier. Similarly, when you need to make decisions around testing, go-live and so on, the suppliers will have seen this completed many times in the past, in your industry, and will have valuable input to these decisions.

In Step 2 – Prepare I spoke about the importance of buyer and seller involvement. Having representation of both on the project board is valuable because they will refer back to the original intentions and benefits the project will deliver.

I am also a big believer in alternative points of view and transparency where possible. Having your organisation and the suppliers represented will ensure you are transparent, it will mean you benefit from a range of inputs and you will ultimately achieve a high-quality outcome. Having an independent experienced member can also help; for example, a WFM delivery industry expert.

Sub-committees/working groups

You may want to include sub-committees or working groups to help clear the path for having some of your decisions accepted. Identify and set up these committees at the outset of your initiative. This will ensure you have the appropriate focus and representation available to work through key items. A common area for a sub-committee or working group is around EBA and rostering.

EBA

As identified in Step 2 – Prepare, in the sections associated with award interpretation, I noted that there are many things to consider associated with the EBA. These issues can be time-consuming and complex to work through. To recap, the key areas for consideration are:

- → configuring directly from an EBA
- → payment to EBA of site practice
- → who makes the interpretation

→ different interpretations of the same EBA
→ what to configure
→ underpayment and overpayment.

Rostering

As identified in Step 2 - Prepare, in the section associated with understanding the rostering models, I noted there are many things to consider associated with the operating models and rostering. These ways of working can be time-consuming and complex to work through. To recap, the key areas for consideration are:

→ types of rostering
→ labour standards
→ roster optimisation; and
→ centralised versus decentralised rostering.

Ensure the sub-committee has processes set up to monitor, inform and make decisions on each of these areas. Also check that there is support from IR and the unions where required.

Methodology

Your organisation, suppliers and delivery team will need to agree on a preferred methodology. Detailed discussion on methodology was provided in Step 3 - Implement. Your organisation may have its own implementation methodology, or you may use one of the traditional methodologies. In addition, your WFM supplier may have its own methodology built from years of delivery experience. Alignment and handover points with the methodologies will be required. For example, will your organisation's methodology reporting method be adopted, or will the supplier's be adopted?

This decision should be made at the outset of the WFM initiative, agreed by the governance committee and understood by all key stakeholders on the project and the implementation team. Making this clear at the outset of your initiative will ensure effective decision-making.

If you do not do this, you run a high risk of your suppliers pulling in different directions, or having differing views about the same topic or definition of an item of work. For example, people with a payroll background will often define a parallel run as comparing historical pay results for an agreed period. People with a WFM background will often define a parallel run as using your new WFM system for the current pay period as well as your legacy system.

Delivery method

Your organisation, suppliers and delivery team will need to agree on a preferred delivery method. This decision should be made at the outset of the WFM initiative, agreed by the governance committee and understood by all key stakeholders on the project and the implementation team. Making this clear at the outset of your initiative will ensure your team is acting and thinking along the same lines.

Detailed discussion on delivery methods - agile or waterfall - was provided in Step 3 - Implement. Will you implement using an agile or waterfall approach? It is key to have these discussions and agree this prior to the commencement of the implementation program. In extreme cases, I have seen a switch from waterfall to agile mid-stream of an implementation. The impacts of this were significant in terms of expectations, skills, competence and cost. The delivery method you choose is entirely up to you, but just be clear on your approach and make sure you have the skills to support the approach.

Scope, plan and timeline

These are perhaps the most fundamental governance items. The scope, plan and timeline should be agreed by the governance committee. Ensure they are understood by all key stakeholders on the project, the implementation team and those people at the operational coal face too. Making this clear at the outset of your initiative will enable everyone to move in the same direction with clarity.

The governance committee and the PMs must be 100 per cent clear about this from the outset. If there is any uncertainty, address it as a priority. Document these areas in a way that everyone can understand. Anyone who is part of the WFM initiative should be able to explain the scope, plan and timeline of your WFM implementation (and they should give the same explanation).

Common questions that arise where these items can be thrown off track include:

- → Which EBAs are in scope?
- → Which rostering methods are in scope?
- → Which sites are in scope?
- → Which business areas are in scope?
- → Which methodology will be used?
- → What is the enterprise architecture?
- → What is the testing process?
- → What business process changes will be adopted?
- → How many deployments will there be?

From a planning perspective, make it clear to your team members what they have to do, when they must do it and what your expectations are of their contribution.

Project risks and issues

Create a culture of acceptance and support around risks and issues. Provide a forum where your business, team and suppliers can raise these items without any fear of blame. Conduct regular reviews and address these items on their merit.

Change control

Change control is inevitable on WFM projects. The key areas where I see change are generally related to which EBAs are in scope, which rostering methods are in scope, which sites are in scope, which business areas are in scope, the number of deployments and your responsibilities versus your product supplier's responsibilities.

It is imperative that your delivery team understand the change control process. This is twofold. Firstly, it involves raising change items.

Secondly, once a change is agreed, you need to ensure that the outcome is communicated back to the team. It is easy for team members to avoid, not follow the process, or fail to adopt the change because they either do not understand it, or they don't know about it.

When considering change, check back to the benefits and business case to ensure the change is in alignment with the agreed organisational goals.

Project costs

Be realistic about your project costs and ensure that the management and governance team have the same understanding of the budget. Are you in alignment with your financial director; is your product supplier working from the same budget you are on?

If you require a cost adjustment, ensure the process for this is understood.

Reporting

Agree on your reporting standards at the outset of the WFM engagement. Consider what methodology or methodologies will be used to provide the reporting templates. Ensure the reporting formats and expectations are understood. Clearly communicate the frequency with which you require your reporting to be completed.

Be as transparent as possible in your project reporting. Keeping your teams informed - from senior management through to those at the coal face - is a key ingredient in keeping your WFM initiative on track.

Suppliers

It is common to have multiple suppliers working across a WFM initiative. Where relevant, introduce your suppliers to each other at the outset of the engagement to increase collaboration.

Suppliers generally work across multiple clients; make sure your priorities are understood by your suppliers and that they are applying the correct resources and focus to make your initiative successful.

Skill levels within the supplier's (and your own) teams will vary so ensure you are getting the appropriate skills for your investment. Sometimes suppliers will swap out staff due to competing priorities (as you will); this will place additional risk on the quality of delivery for your WFM initiative.

Understand at the start of your WFM initiative what value-add your supplier provides in areas such as account management, support, knowledge forums and so on. Keep your product and service suppliers accountable for their responsibilities.

Having a relationship in place with your supplier that separates the delivery and commercial negotiation can remove tensions that sometimes arise if the delivery person is having commercial discussions.

Go-live

WFM projects deal with your people, the way they get paid and your operating models. These are key considerations when you make any go-live decisions. Your options for go-live are typically phased or big-bang.

Phased

When you phase go-live, a section of your business is selected and it goes live at a given time. Take into consideration things such as support of legacy systems, pay periods, end of month, end of year, blackout periods, or other project/business initiatives.

It is difficult to separate rostering from EBA/pay results when going live, as rostering decisions have a direct impact on pay. For example, the times scheduled to start and stop work have a direct impact on the number of overtime hours. It's important to realise that your way of paying your employees and way of operating will change at the same time with the introduction of automated rostering and payment.

It is possible to separate shift-based rostering from roster optimisation, i.e. you can go-live initially with shift-based rostering and then implement roster optimisation later.

Big-bang

A big-bang go-live is where all of your people, process and technology changes are adopted at the same time. Generally, big-bang go-lives are the result of a compelling business need. For example, you may face significant contractual penalties for being on an existing system or you may be close to end-of-life for a legacy software solution.

My experiences of big-bang, especially at the large customer enterprise level, have not been positive, particularly for the operational user. Processes were not bedded down, payments were wrong, training was lacking, support structures were deficient. I am not saying don't adopt a big-bang go-live approach but, if you do, you need to have a high degree of diligence and governance in place to mitigate the risks associated with this method.

Support structure

WFM initiatives have a large footprint with your operational workforce. Consider your go-live and ongoing support structures well in advance to make sure you provide adequate support for your people. The larger the workforce, the more emphasis is required.

Refer also to the considerations in Chapter 9, Step 3 – Implement, associated with change and go-live support.

Business impacts

Before we leave the discussion on tracking your implementation, let's consider the business impacts.

Business engagement, change and adoption

No matter how good your technical solution, the people in your business will determine the success or failure of your WFM initiative. If the users do not adopt the solution, the WFM initiative will not be successful.

Ensure you have representation across all your business areas and that you have a process in place to solicit objective feedback that feeds up to the governance level.

People assignment

Earlier in Step 2 - Prepare we looked at people requirements from a skills and timing perspective. Keep monitoring this throughout the course of the project. Not having the correct allocation of staff will lead to unnecessary stress and impede the quality of the solution. In some cases, I have seen this result in project team churn.

You need to answer two simple questions:

1. Are the people on the project happy?
2. Are the people in my business happy?

Top take outs

→ The items raised in this chapter should complement your preferred governance model.

→ Many WFM nuances are important to consider at a governance level to mitigate risk and increase organisational adoption.

→ Set up sub-committees early in the engagement to deal with areas of high impact business change. Common areas are EBA and rostering.

→ Suppliers have substantial industry and product knowledge; their input to organisational governance is highly valuable.

→ WFM initiatives have a big effect on the operational areas of the workforce. Conduct appropriate due diligence to determine the most applicable go-live approach and support the workforce during the process.

Where to next?

Step 4 - Track focused on the identification of a number of areas to help keep your WFM initiative on track. The next chapter and final step in the methodology, Step 5 - Measure, focuses on the measures you need to take to ensure you receive value from WFM for the long term.

Chapter 11

STEP 5 – MEASURE

It's been my pleasure to see many organisations deliver WFM initiatives successfully. But, how do you measure success? On some occasions the initiative delivers what senior management is expecting and on other occasions the initiative misses the expected outcome. Generally, in the sales cycle there is a clear focus of what the initiative is required to deliver; clear goals and expectations of benefits are set. As the initiative moves on through the implementation process there is an increased likelihood that the desired outcome pivots. Generally, the expected results from senior management are not readily visible to the project team and there is a misalignment of views on how to track the benefits.

How to achieve optimal value from WFM

In this final step, we focus on continually reviewing the benefits to ensure you receive optimal value from your WFM initiatives.

There is a strong relationship between the align and measure steps of the methodology. When you are at the align stage, it is an ideal time to capture the benefits of your WFM initiative, paying consideration to the measures you require. This will allow clear focus on delivery of outcomes. Remember the old adage: 'You can't manage if you don't measure'.

Figure 11.1: Smart WFM Methodology: Step 5 – Measure

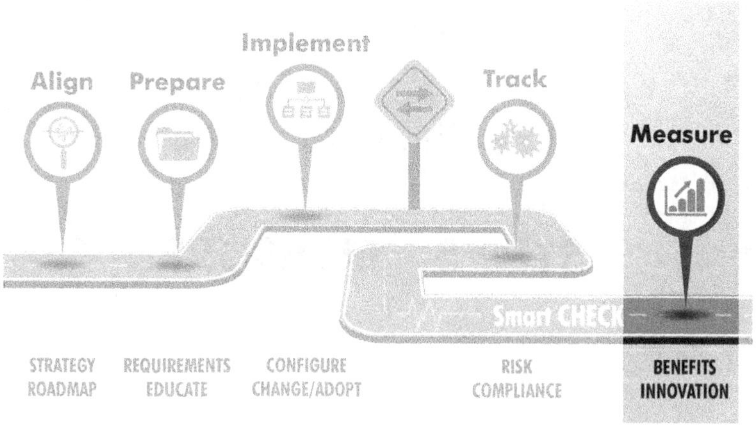

Benefit measurement framework

Perhaps one of the greatest challenges with measurement is knowing how and what to measure, so here I share a few tips that have worked for me.

Making available your desired outcomes throughout the organisation, from senior management to the operational coal face, will enable you to align and measure benefits. You will also need to understand the organisational goals by defining them in terms of workforce goals, align them to WFM benefits and, finally, determine what needs to be enabled from a WFM perspective to achieve this.

By taking this holistic approach, if someone asks why you enabled a certain WFM technology function or process, you can refer back directly to an organisational goal.

I've created a four-stage benefit measurement framework that you can use to structure your thinking and enable measures to be established (see Figure 11.2). The framework is introduced by way of an example. There may be one or many relationships across the goals, benefits, enablers and measures; there may be more than one workforce goal to achieve an organisational goal; there may be more than one WFM enabler to achieve a WFM benefit.

Figure 11.2: Smart WFM benefit measurement framework

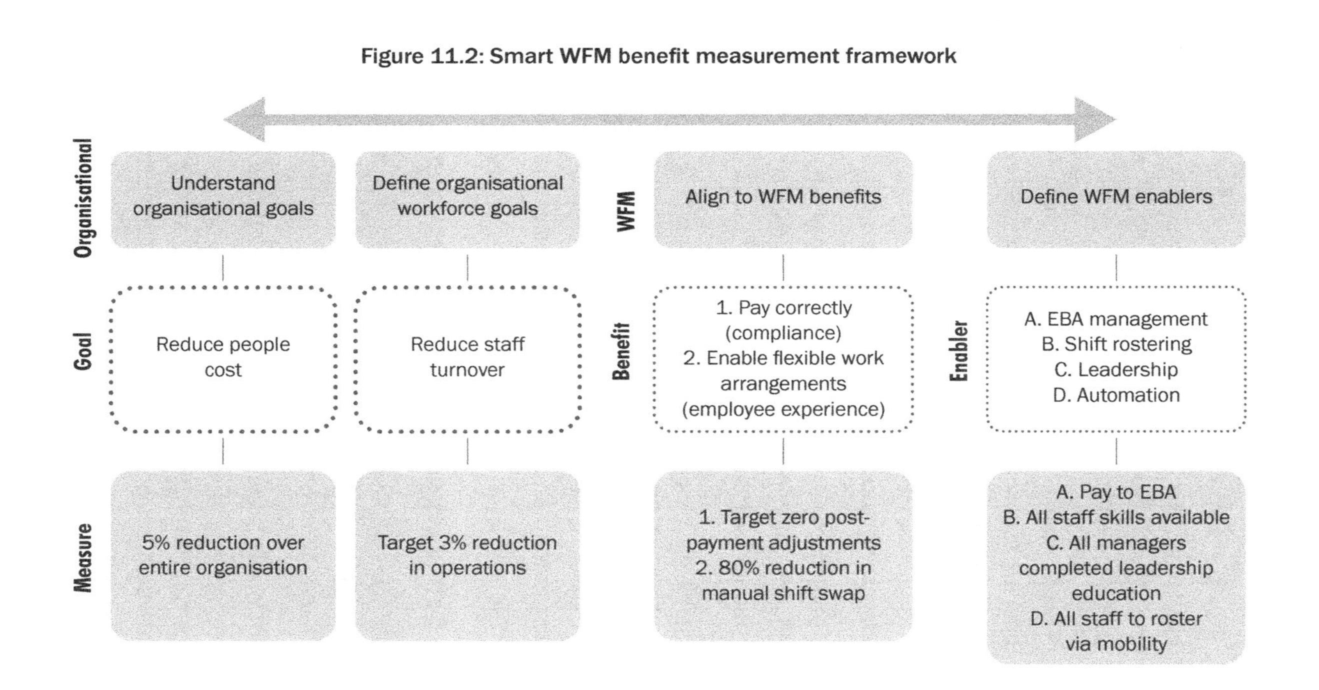

This objective of the framework is to take it and tailor it to your organisation's specific needs.

Note: You may see some similarities to the traceability discussion I introduced earlier in Step 3 - Implement. A traceability matrix usually goes to the next level of detail - individual functional requirements - and does not always define the measures.

In this framework, stages 1 and 2 are organisationally focused, while stages 3 and 4 are specifically WFM focused. These stages do not necessarily need to be completed in a sequential manner, but all should be completed to provide the whole picture.

Stage 1 – Understand organisational goals

The first WFM fundamental is to understand the organisational goals. If you embark on a WFM initiative, it's crucial to understand what business goals you are addressing.

Organisational goals will often be set across all your business areas, e.g. HR, finance, sales and logistics. In the example above, the organisational goal is to reduce people costs by 5 per cent across the entire organisation, therefore you would need to understand how this figure plays out across all the areas combined and how it relates to each business area.

Stage 2 – Define organisational workforce goals

The alignment of organisational goals to WFM was introduced in Step 1 - Align. It's your job to unpack these organisational goals and how they relate to your workforce.

In the example, the organisation has determined that a 3 per cent reduction in operational staff turnover will deliver the required cost savings that contribute to the required people savings. The cost of employee turnover is also a measurable calculation that is used to determine the 3 per cent cost saving.

Stage 3 – Align to WFM benefits

The benefits of WFM were introduced in Step 1 - Align. These are summarised below:

- → real-time visibility of people and costs
- → compliance
- → accuracy
- → full visibility of labour force
- → forecasting
- → efficiency
- → productivity
- → culture and purpose
- → employee experience and engagement
- → customer experience and engagement
- → increased sales/service output
- → quality
- → leadership.

All these are high-level areas where benefits can be obtained. Within each of these areas, there are many ways to achieve the high-level outcome. For example, in the example to reduce staff turnover, the organisation has determined there are two key benefits from a WFM perspective to achieve this:

- → pay staff correctly (a compliance benefit)
- → enable flexible work arrangements (an employee experience benefit).

There are now tangible measures that can be applied to achieve these benefits:

- → target zero post-payment adjustments
- → an 80 per cent reduction in manual shift swaps.

Stage 4 – Define WFM enablers

The fourth and final stage of the benefit measurement framework is to define the WFM enablers. These enablers are at the lowest level of measurement granularity. Think of these areas as being broadly related to people, process and technology. How are you going to achieve your goals and benefits?

Continuing with our example to pay people correctly and enable flexible working arrangements requires a number of WFM enablers, which are defined as:

→ EBA management
→ shift rostering
→ WFM leadership
→ WFM automation.

There are now tangible measures that can, in turn, be applied to achieve each of these benefits:

→ pay to EBA
→ all staff skills available
→ all managers completed leadership education
→ all staff to roster via mobility.

WFM maturity measurement

Over time, you will be looking to improve your organisation's WFM maturity. Generally, the more mature your organisation (in terms of WFM), the greater the benefits you can achieve from WFM. In Step 1 – Align, we mapped WFM maturity to the following progression:

1. Time recording.
2. Award interpretation.
3. Rostering.
4. Forecasting/work based/route.
5. Optimisation.

It is possible to achieve roster optimisation from the outset of a WFM initiative, but you will need to simultaneously adopt the appropriate enablers for time recording, award interpretation, rostering and forecasting/work based/route. Careful preparation will be required to achieve this.

To make this tangible, let's look back at the example provided in the benefit measurement framework. Staff skills are required to enable shift-based rostering. Now, to complete roster optimisation, you require skills, labour standards and consideration of additional business rules.

It is possible to enable roster optimisation at the outset of your initiative. Given the amount of effort to implement and the change required to adopt, it may make more sense for your organisation to define and bed in skills rostering prior to moving to roster optimisation.

Smart CHECK framework

You will recall I introduced Smart CHECK when I defined the Smart WFM Methodology in Chapter 6. At any time during the business lifecycle, a Smart CHECK can be completed to mark where you are at and provide recommendations on what you need to do next. The Smart CHECK can be completed by step, by specific tasks in a step, or any combination of step and task.

I noted in my introduction of Smart CHECK that it generally looks at people, process, technology and/or financials and delivers a gap analysis. To expand on this, you may want to consider additional areas such as:

- → scope of area to be assessed
- → assessment and recommendation
- → issues and risks
- → opportunities for improvement/gaps
- → barriers to achieving good practice
- → evidence.

When assessing using Smart CHECK, also consider the time being spent on strategic, operational and administrative tasks, both now and in future state, in terms of achieving good practice. Perhaps you will need to rebalance the time allocation to achieve your desired outcome.

As the end of this book you will see that you are able to go online and complete a readiness assessment that benchmarks the ability to achieve WFM outcomes in your business.

Continuous improvement and innovation

Much has been written about both these concepts; if you do an internet search you will find a significant amount of literature. I therefore have only a couple of comments based on my WFM experiences.

With the growing prominence of cloud and significant technological advancements associated with AI, data science, machine learning, deep learning, IoT and omni channel it is important to stay current.

Ensure you have people driving continuous improvement and innovation. You need to continually relate this to your organisational goals, workforce goals, WFM benefits and WFM enablers. Take advantage of improvements, adopting the ones that are right to meet your organisation's needs.

Top take outs

→ Use a consistent framework to measure your business outcomes.

→ Workforce maturity will play a large part in how quickly you can adopt change.

→ Use a Smart CHECK at any time during your initiative to mark where you are at and what you need to do next.

→ Develop a culture of continuous improvement and innovation.

Where to next?

This chapter introduced the final step in the Smart WFM 5-Step Methodology to ensure the business benefits of your organisation's WFM initiative are met and that checks and balances are put in place to gain ongoing benefits from WFM.

The next step is to put the learnings and experiences from this book into practice. I wish you luck with your people-based WFM initiatives.

CONCLUSION

I hope this book has succeeded in achieving some of the objectives I had when I decided to write it, which as you may recall were:

1. To teach organisations what WFM is by examining the various areas of WFM through a business and future-focused lens.
2. To make organisations aware of key, predictable WFM-related issues that arise.
3. To present my proven 5-Step Smart WFM Methodology that can be used as a template to overcome WFM-related issues and deliver long-term, effective workforce value.

It's up to you now to take these learnings and apply them in ways that make sense to achieve your organisation's WFM outcome. As our knowledge grows, our ability to manage our workforce more effectively will also grow.

I trust that as we network in this digital age and continue to learn, the business, employee, customer and social outcomes continue to become greater.

I'll leave you with a quote from Walt Disney:

> *'Our greatest natural resource is the minds of our children.'*

LIST OF ABBREVIATIONS

3D	Three Dimensional
AI	Artificial Intelligence
API	Application Programming Interfaces
BOTS	Robots
BYOD	Bring Your Own Device
EBA	Enterprise Bargaining Agreement
ERP	Enterprise Resource Planning
ESS	Employee Self Service
HCM	Human Capital Management
HR	Human Resources
IoT	Internet of Things
IaaS	Infrastructure as a Service
IPS	Implementation Planning Strategy
IR	Industrial Relations
ISO	International Organization for Standardization
IT	Information Technology
KPIs	Key Performance Indicators
IPS	Implementation Planning Study
MSS	Manager Self Service
MSA	Master Service Agreement
OH&S	Occupational Health and Safety
NDIS	National Disability Insurance Scheme
PaaS	Platforms as a Service
PDR	Personal Data Repository
PEPM	Per Employee Per Month
PM	Project Manager
PoE	Power over Ethernet
POS	Point of Sale
RACI	Responsible, Accountable, Consulted, Informed
R&D	Research and Development

RFI	Request for Information
RFP	Request for Proposal
RFID	Radio Frequency Identification
RPA	Robotic Process Automation
SaaS	Software as a Service
SMB	Small-to-medium business
SOW	Statement of Work
WFM	Workforce Management

ACKNOWLEDGMENTS

There are many friends, partners and colleagues around me who I have learned from; they have supported the ideas you find in this book and put trust in me to write it. Thank you to everyone for enabling this to happen. I have thoroughly enjoyed the writing process and can't put into words how personally satisfying it was to write this book. I feel blessed and privileged to have written this book for you.

As part of my ongoing education and desire to learn, I have read countless books over the last 18 months. I'd like to recommend a few of these, not all of which are related to WFM but are related to running a digital business that has a workforce:

→ *HR from Now to Next: Reimagining the workplace of tomorrow.* Jason Averbook. © 2014 Jason Averbook

→ *Life in Half a Second: How to achieve success before it's too late.* Matthew Michalewicz. © 2013 Matthew Michalewicz. Hybrid Publishers Pty Ltd

→ *Rebirth of the Salesman: The World of Sales Is Evolving. Are You?* Cian McLoughlin. © 2015 Cian McLoughlin. OMNE Publishing

→ *Key Person of Influence: The Five-Step Method to become one of the most highly valued and highly paid people in your industry.* Daniel Priestly. © 2014 Daniel Priestly. Rethink Press

→ *The Ultimate Guide to Remote Work: How to Grow, Manage and Work with Remote Teams (Zapier App Guides Book 3).* Wade Foster, Danny Schreiber, Alison Groves, Matthew Guay, Jeremy DuVall, Belle Cooper. © 2015 Zapier Inc.

→ *The Network Imperative: How to Survive and Grow in the Age of Digital Business Models.* Barry Libert, Megan Beck, and Yoram Wind. © 2016. *Harvard Business Review*

→ *Chapter One: You have the power to change stuff.* Daniel Flynn. © 2016 Daniel Flynn. The Messenger Group Pty Ltd.

→ *The Fourth Industrial Revolution*. Klaus Schwab. © World Economic Forum, 2016

→ *No Boundaries: How to use Time and Labor Management Technology to Win the Race for Profits and Productivity.* Lisa Disselkamp. © 2009 Lisa Disselkamp. John Wiley & Sons

→ *Workforce Asset Management Book of Knowledge.* Editor: Lisa Disselkamp. March 2013. John Wiley & Sons

Thanks to the authors of the countless blogs, Twitter messages and podcasts I have read and listened to. I would like to note a few, in particular:

→ **The Firing Line:** Bill Kutik, https://www.youtube.com/channel/UC4UTOIN_lzmNxKewHK7NCtA

→ **HR Happy Hour:** Steve Boese and Trish McFarlane, https://itunes.apple.com/au/podcast/hr-happy-hour/id325399068?mt=2

→ **Talking People and Tech:** David Guazzarotto and Jared Cameron, https://itunes.apple.com/nz/podcast/talking-people-and-tech-brought-to-you-by-future-knowledge/id1126201501?mt=2

There are three thought leaders in the space for whom I hold the highest regard: Jason Averbook, Josh Bersin and Rob Scott – love the way you push the boundaries and keep us moving forward!

REFERENCES

Aljaber, T. 'The iron triangle of planning: The ultimate balancing act and how to achieve agile project management nirvana', Atlassian, viewed [2 December 2017], <https://www.atlassian.com/agile/agile-iron-triangle>.

Fair Work Commission, *Awards & Agreements*, viewed [2 August 2017], <https://www.fwc.gov.au/awards-and-agreements/agreements>.

Garton, E. 2017, 'What If Companies Managed People as Carefully as They Manage Money?', *Harvard Business Review*, 24 May, viewed 12 July 2017, <https://hbr.org/product/what-if-companies-managed-people-as-carefully-as-they-manage-money/H03ODJ-PDF-ENG?referral=03069>.

Google Patents, *Bundy Clock*, viewed 31 July 2017, <https://www.google.com/patents/US452894>.

Markovski, M. 2018, Apps Run the World, Top 10 Workforce Management Software Vendors, Market Forecast 2016–2021, and Customer Wins', 8 January, viewed 17 January 2018, <https://www.appsruntheworld.com/top-10-workforce-management-vendors-market-forecast-2016-2021-and-customer-wins/S/>.

NDIS, *Home Page*, viewed 5 August 2017, <https://www.ndis.gov.au>.

Schwab, K. 2016, *The Fourth Industrial Revolution*, Penguin Books Ltd, London, UK.

Ventana Research, 'Workforce Management Value Index 2017, Vendor and Product Assessment', January, viewed 10 February 2017, <https://www.ventanaresearch.com/value_index/human_capital_management/workforce_management>.

INDEX

R

S

W

'We must think globally, while acting locally to embrace clients, family, community and the environment. This will allow business to reach new heights, increasing value within the workplaces of tomorrow.'

Jarrod McGrath
Founder and CEO – Smart WFM

smartwfm.com

Make the Pledge

Smart WFM is proud of its relationship with Pledge 1%. We embedded corporate philanthropy into our business model from inception. Providing a greater sense of purpose to why we do what we do is important to us.

We pledge to the Cathy Freeman Foundation and support its commitment to bridging the education gap between Indigenous and non-Indigenous Australians.

Part of the sales revenue from this book will be donated to Pledge 1% and the Cathy Freeman Foundation.

Find out more:

PLEDGE 1%

pledge1percent.org

cathyfreemanfoundation.org.au

Our Vision

Empower the workforce now and into the future

Our Mission

Maximise people value, productivity and experience

Our Core Values

- Make a positive social contribution
- Enable a greater sense of purpose for our team
- Make people's lives better
- Harness and share knowledge
- Act with honesty and integrity
- Think from a customer perspective

Who we are

- Industry pioneer
- Easy to work with
- Consulting – advisory and implementation
- Business focused
- Product supplier, independent
- Built upon experience and reputation

How we do it

- Future thinking
- Big picture focus
- Simplified approach
- Strategic alignment and partnering
- Proven Smart WFM 5-Step Methodology
- Embrace digital
- Data driven decisions
- Measurable value